CAMBRIDGE LIBRARY COLLECT

Books of enduring scholarly value

Travel and Exploration

The history of travel writing dates back to the Bible, Caesar, the Vikings and the Crusaders, and its many themes include war, trade, science and recreation. Explorers from Columbus to Cook charted lands not previously visited by Western travellers, and were followed by merchants, missionaries, and colonists, who wrote accounts of their experiences. The development of steam power in the nineteenth century provided opportunities for increasing numbers of 'ordinary' people to travel further, more economically, and more safely, and resulted in great enthusiasm for travel writing among the reading public. Works included in this series range from first-hand descriptions of previously unrecorded places, to literary accounts of the strange habits of foreigners, to examples of the burgeoning numbers of guidebooks produced to satisfy the needs of a new kind of traveller - the tourist.

A Voyage to the South Atlantic and Round Cape Horn into the Pacific Ocean

British naval officer James Colnett (1753–1806) served on many voyages during his career. He was a midshipman on Captain Cook's second voyage, and in 1774, he was first to sight New Caledonia, which led to Cook naming Cape Colnett after him. Later in his career, he was in command of the fur-trading expedition that resulted in the Nootka Crisis and near-war between Spain and England. In this book, first published in 1798, Colnett gives an account of the voyage he commanded to the Antarctic in 1793. The expedition's success at charting suitable places for ships to anchor was instrumental to the development of the whaling industry in the area. This book contains accounts and maps of the islands and coastlands visited during the expedition. Among the islands visited were the Galapagos Islands, and Charles Darwin is known to have had a copy of this book on HMS *Beagle*.

Cambridge University Press has long been a pioneer in the reissuing of out-of-print titles from its own backlist, producing digital reprints of books that are still sought after by scholars and students but could not be reprinted economically using traditional technology. The Cambridge Library Collection extends this activity to a wider range of books which are still of importance to researchers and professionals, either for the source material they contain, or as landmarks in the history of their academic discipline.

Drawing from the world-renowned collections in the Cambridge University Library and other partner libraries, and guided by the advice of experts in each subject area, Cambridge University Press is using state-of-the-art scanning machines in its own Printing House to capture the content of each book selected for inclusion. The files are processed to give a consistently clear, crisp image, and the books finished to the high quality standard for which the Press is recognised around the world. The latest print-on-demand technology ensures that the books will remain available indefinitely, and that orders for single or multiple copies can quickly be supplied.

The Cambridge Library Collection brings back to life books of enduring scholarly value (including out-of-copyright works originally issued by other publishers) across a wide range of disciplines in the humanities and social sciences and in science and technology.

A Voyage to the South Atlantic and Round Cape Horn into the Pacific Ocean

For the Purpose of Extending the Spermaceti Whale Fisheries, and Other Objects of Commerce

JAMES COLNETT

CAMBRIDGE
UNIVERSITY PRESS

CAMBRIDGE UNIVERSITY PRESS
Cambridge, New York, Melbourne, Madrid, Cape Town,
Singapore, São Paolo, Delhi, Mexico City

Published in the United States of America by Cambridge University Press, New York

www.cambridge.org
Information on this title: www.cambridge.org/9781108048354

© in this compilation Cambridge University Press 2012

This edition first published 1798
This digitally printed version 2012

ISBN 978-1-108-04835-4 Paperback

This book reproduces the text of the original edition. The content and language reflect the beliefs, practices and terminology of their time, and have not been updated.

Cambridge University Press wishes to make clear that the book, unless originally published by Cambridge, is not being republished by, in association or collaboration with, or with the endorsement or approval of, the original publisher or its successors in title.

A VOYAGE

TO THE

SOUTH ATLANTIC

AND ROUND

CAPE HORN

INTO THE

PACIFIC OCEAN,

FOR THE PURPOSE OF EXTENDING THE

SPERMACETI WHALE FISHERIES,

AND OTHER OBJECTS OF COMMERCE,

BY ASCERTAINING

THE PORTS, BAYS, HARBOURS, AND ANCHORING BIRTHS,

IN CERTAIN ISLANDS AND COASTS IN THOSE SEAS

AT WHICH THE SHIPS OF THE BRITISH MERCHANTS MIGHT BE REFITTED.

UNDERTAKEN AND PERFORMED

BY CAPTAIN JAMES COLNETT,

OF THE ROYAL NAVY, IN THE SHIP RATTLER.

LONDON:

PRINTED FOR THE AUTHOR,

BY W. BENNETT, MARSHAM STREET, WESTMINSTER.

SOLD BY A. ARROWSMITH, CHARLES STREET, SOHO; STOCKDALE,
PICCADILLY; EDGERTON, CHARING CROSS, ELMSLY,
STRAND; AND WHITE, FLEET STREET.

1798.

CONTENTS.

	Page.
Dedication	1 *to* 4
Introduction	1 - 18

CHAPTER I.
Passage of the Rattler from England to Rio Janeiro   1 - 7

CHAPTER II.
Attempt to discover Isle Grand; and Passage Round Cape Horn   8 - 18

CHAPTER III.
Remarks on the Navigation round Cape Horn   19 - 20

CHAPTER IV.
Route from Cape Horn to making the Coast of Chili, and the Isles St. Felix and St. Ambrose   21 - 37

CHAPTER V.
Route of the Rattler from the Isles Saint Felix and Saint Ambrose, to the Coast of Peru   38 - 46

CHAPTER VI.
The Galapagoe Isles   47 - 61

Passage

CHAPTER VII.

Passage from the Galapagoe Isles, to Isle Cocas   62 to 74

CHAPTER VIII.

Route from Isle Cocas, to the Coast of Mexico; and Isles Santo Berto, and Rocka Partido, from thence to the Coast of California, and account of our cruize in the Gulf of California, with our return to Socoro after searching for Isle St. Thomas - - - - - 75 - 121

CHAPTER IX.

The Rattler quits the Isle of Socoro for the Coast of Mexico: some account of our transactions there, and while we lay at anchor before the Island of Quibo, in the Gulf of Panama, to our arrival at the Isles of the Galapagoes, on and near the Equator - - - - - 122 - 160

CHAPTER X.

The Rattler leaves the Galapagoe Isles and Coast of Peru, for the Isles Saint Felix and Saint Ambrose, on the Coast of Chili: from thence she rounds Cape Horn, on her Passage to Isle Saint Helena, in the Atlantic Ocean - - 161 - 176

CHAPTER XI.

From Isle Saint Helena to England - - 177 - 179

TO

SIR PHILIP STEPHENS, Bar.t M.P.

TO

Sir PHILIP STEPHENS, Bart.

ONE OF THE

LORDS COMMISSIONERS OF THE ADMIRALTY,

MEMBER OF PARLIAMENT FOR SANDWICH,

F. R. S. &c. &c. &c.

Sir,

In dedicating to you the voyage, which is the subject of the following pages, my gratification would indeed be sincere, (did the work possess an importance which might fully claim your regard:) humble, however, as it's

pre-

pretenfions are, the opportunity, it gives, muft ferve to teftify my fincere refpect, my lafting gratitude, for your favors; and I repofe with fome fecurity, on an extenfion, of that protection, towards this volume, which has fo often been the encouragement, of my profeffional exertions.

The difficulties which navigators have experienced, in traverfing the South Seas and Pacific Ocean, have evidently, from the commencement of your connection with the Admiralty, excited your particular confideration: and it is certain, in all the changes to which that Board has been fubject, that the explorer of remote feas, seas, has found in you, a zealous advocate to represent his claims to those, with whom it remained, to bestow the reward due to his endeavours and merits.

As far as I am individually concerned, it is with sincerity I aver, that in all situations of trial and difficulty on unknown and unfriendly coasts, I have found an unceasing consolation in the hope that I possessed your good opinion; and that in the end, my services would obtain, through the channel of your faithful explanation, a liberal requital.

If any distinct praise can confer an honor on your name, (beyond what it bears for the general ability, zeal and integrity,

integrity, which have been invariably manifested, during the extensive period of your official services), it is most assuredly due for your public, as well as *private* acts of friendship to those, who like me, have embarked for the purpose of enlarging the bounds of Navigation and Commerce; and I feel a decided conviction, that every follower of the able Captain Cook, will give a cordial assent to this tribute.

   I am,
    Sir,
     Your truly obliged
      and most obedient
       humble Servant,

No. 17, Milman Street, Bedford Row.

    JAMES COLNETT.

INTRODUCTION.

It will not, I prefume, be confidered as altogether unneceffary or uninterefting, if I offer to the public attention a fummary account of my voyages previous to that which is the fubject of this volume.

I had been already engaged in various commercial undertakings on the North-Weft coaft of America, during a period of feven years: But I never ceafed to blend the zeal of my naval character with the fpirit of commercial enterprize,

INTRODUCTION.

and accordingly searched the coast from 36° to 60° North; the inland part of which was before little known to European navigators. I also surveyed several bays, harbours and creeks, and discovered many considerable inlets, particularly between 50° and 53° North, which were supposed to communicate with Hudson's Bay*. I likewise made two voyages to China, but, on my return from the first of them, was unfortunately captured by the Spaniards at Nootka-Sound, and sent a prisoner to the port of St. Blas in the Gulph of California. From thence I was permitted to go to Mexico, to make my appeal to the Viceroy; a journey, including my return to St. Blas, of eighteen hundred miles. But after being detained as a prisoner thirteen months, and having lost four of my five vessels, with most of my Officers and half the crew, who had become the victims of disease, I was at length indebted

---

* These inlets have since been more particularly examined by Capt. Vancouver of the Royal Navy, by order of the Board of Admiralty.

indebted for my liberty to the spirited conduct of the Court of Great-Britain, as soon as it was officially informed of the insult offered to her colours, of the piracy committed on her merchants, and the cruelty exercised towards her subjects.

I now returned to Nootka, in the only vessel which remained to me; and, after suffering incredible hardships from a want of provisions, and the ship getting several times on shore, I procured another valuable cargo of furs and proceeded to China. A prohibition, however, of the sale of these articles. having taken place at that port during my absence, I did not remain there, but, in a short time, set sail, and, at the request of those gentlemen who were joint agents with me, coasted for a market to the West side of Japan, and East side of Corea; a coast which had never before been visited by an European vessel. Here an encouraging prospect of a new and valuable commerce for my country unfolded itself

itſelf before me; when, in a typhoon, in the Latitude of 38° North, on the coaſt of Corea, I was ſo unfortunate as to loſe my rudder, which obliged me to put back into the port of Chuſan in the Northern parts of China, where my loſs was repaired, as well as could be contrived, without the profeſſional aid of a carpenter. At this place, I was, by no means, well treated by the Chineſe: indeed, to prevent the being plundered of our cargo, and to ſave ourſelves from the ill-uſage which we might perſonally ſuffer, if we ſhould fall into their hands, we cut from our anchor, and, baffling the purſuit of thirty-ſix armed junks, returned to Canton* Here I was threatened with ſtill greater evils, for being obliged from the wretched ſtate of the ſhip, which was almoſt a wreck, to paſs the Bocca Tigris, without ſtopping to obtain the neceſſary paſſport, the

---

* I communicated the track of this voyage to Mr. Arrowſmith, Geographer, and it may be ſeen in his chart of the Pacific Ocean which will be ſoon publiſhed.

the Chinese made this act of necessity on my part, an official pretext on theirs to seize the vessel and cargo; and having contrived by stratagem, to get me out of her, detained me as a prisoner in Canton during five days: But when the various unfortunate circumstances which had befallen me were explained, and the truth of them duly attested, before the judicial Mandarins, I was ordered to be enlarged, while my vessel was sent for me, at the expence of the Chinese Government, to Macao to be sold, and myself and cargo, by the same superior interposition, put on board the East-India Company's ship, General Coote, bound for England. On my arrival there, the India Company purchased my cargo for nine thousand seven hundred and sixty pounds sterling. A full account of this voyage and the circumstances connected with it, together with the charts and drawings made to illustrate it, were left in England when I proceeded on my last voyage; a narrative of which, as well as of that I made to Japan

and

and Corea, will hereafter, I truſt, be communicated to the public.

In a ſhort time after my arrival in England, an application was made to me by different merchants to undertake another voyage to the countries I had lately viſited, on a ſalary of ſix hundred pounds per ann and the reimburſement of all my expences, which were the conditions of my former voyages, or to go on my own terms.

So long an abſence from my country as a period of ſeven years, had occaſioned a very conſiderable change in my ſituation. During that interval, death had deprived me of my neareſt relations: many of thoſe friends alſo whom I left in England, when I departed from it, and whom I hoped to ſee again on my return to it, were now no more; ſo that few or no objects were left to check or interrupt the honeſt ambition I had never ceaſed to poſſeſs of riſing in the Royal Navy,
which

which was my original and favourite profeffion. I had ferved on the Quarter-deck of a man of war from the year 1769, and performed the fecond circumnavigating voyage with Capt. Cooke as a midfhipman, on board the Refolution: I had alfo enjoyed for fixteen years the rank of a lieutenant. I accordingly addreffed a letter to the Right Honourable the Lords Commiffioners of the Admiralty, in which, after fetting forth my fervices and fituation, I requefted their Lordfhips, if the profeffional exertions of my paft life did not appear, in their opinion, to merit profeffional promotion, or if they had no employment to offer me in order to obtain it, that they would be pleafed to grant me permiffion to re-engage in my commercial purfuits. Captain Marfhall, one of the Commiffioners of the Victualling Office, under whofe command I had the honour to ferve, during feveral years, in the rank of firft lieutenant, fupported me in this application, which was effectually forwarded by Mr. Stephens; who

was

was uninfluenced by any claim in my favour, but such as my services, whatever they might have been, suggested to the justice and generosity of his character.

In consequence of this application, I was informed that the Board of Admiralty had nominated me to undertake a voyage, planned in consequence of a memorial from merchants of the City of London, concerned in the South Sea Fisheries, to the Board of Trade; for the purpose of discovering such parts for the South Whale Fishers who voyage round Cape Horn, as might afford them the necessary advantages of refreshment and security to refit.

This memorial stated the calamitous situation of the ship's crews employed in this trade, from the scurvy and other diseases, incident to those who are obliged to keep the seas, from the want of that relief and refreshment, which is afforded by intermediate harbours.

The

The Spaniards, it is true, had, of late, admitted ships into their ports for the purpose of refitting; but, from the latest accounts received before I sailed from England, this permission was so restricted as to amount almost to a prohibition, in which it was continually expected to end. It became therefore an object of great importance to obtain such a situation as our commerce required, independant of the Spaniards; as it would, in a great measure, lessen their jealousy, and, at the same time, accomplish the sanguine wishes of our merchants. Impelled by these views and interests, those gentlemen made a proposal to Government to carry out an Officer in one of their ships, in order to discover such a situation, for a gratuity of 500*l*. sterling. This proposal was accordingly accepted, and Messrs. Enderby and Sons, merchants of great property as well as commercial character in London, and who possessed the largest concern in this branch of the Whale Fishery, were pleased to express their satisfaction at my being named to conduct the intended voyage.

There being at this time, no private vessel in the Thames for sale, which was properly constructed for the purposes of this expedition, a proposition was made to the Lords Commissioners of the Admiralty to lend one of his Majesty's small ships of war for the service, to be returned in the same state, at the conclusion of the enterprize. This plan was favoured with their Lordships approbation; and several vessels being proposed and examined with this view, the Rattler sloop of three hundred and seventy-four tons burthen, was selected, then laying at his Majesty's yard at Woolwich, for considerable repairs. It was, however, found, on more mature consideration, that the alterations necessary to be made for the whaling trade, would render her unfit for his Majesty's future service. An offer was therefore made to the Lords of the Admiralty

Admiralty to purchase the said sloop; and their Lordships thought proper to favour it with their acquiescence; a proof of their indulgent attention to any proposal that may tend to advance the interests or extend the limits of the British commerce, and fisheries An official order was accordingly given for the delivery of the sloop, on the purchase-money being paid, according to her valuation; and, on the following day, she was received from the Officers of the King's yard, and transported to Mr. Perry's dock, at Blackwall, in order to be repaired and fitted for the particular service in which she was to be employed.

Of this vessel I was appointed sole Commander, which, however, was a mere complimentary nomination, as no pecuniary advantage whatever was attached to it; exclusive of what I might derive from the subsequent generosity of Government. A whaling master and a crew, consisting of twenty-five perſons, men and boys, were alſo appointed; and with the advice and aſſiſtance of my kinſman Mr. Binmer, firſt aſſiſtant ſurveyor of the Navy, who had ſuperintended the original conſtruction of the ſloop, ſuch alterations were made, as, without diminiſhing her ſtrength, or leſſening her powers of ſailing, were perfectly adapted to the commercial undertaking in which ſhe was about to be employed: ſo that ſhe was equipped and ready to leave the river by the eleventh day of November, 1792. Every nautical inſtrument, neceſſary for determining the longitude and making other uſeful obſervations, were alſo put on board: and I purchaſed of Meſſrs. Enderby's one half of the ſhip, which muſt at once have checked any apprehenſion on their part, that their private concern would be interrupted or receive any detriment from the attention I might pay to ſuch inſtructions as were communicated to me by Mr. Stephens.

Meſſrs.

Meffrs. Enderby and Sons had fitted out the ſhip: but neverthelefs, I ſpared no expence in providing myſelf with all things which my experience of long voyages, as well as my knowledge of the ſeas, I was preparing once more to traverſe, ſuggeſted to me as neceſſary for preſerving the health of thoſe who were to ſail with me. I alſo purchaſed the various voyages of former navigators, and ſuch books on the ſubjects of natural hiſtory, as might aſſiſt me in my purſuits, and enable me to furniſh inſtruction in thoſe branches of ſcience connected with my own; and which remote navigation might tend to advance. In ſhort, I determined to ſpare no exertion in fulfiling every object of the voyage, which had been entruſted to my care and direction.

As my inſtructions were not ready by the eleventh of November, the ſhip was ordered round to Portſmouth, to wait for me. But while I was in daily expectation of receiving my diſpatches, the
unexpected

unexpected and alarming conduct of the French Nation, demanded the whole attention of Government, and occasioned an unavoidable suspension of my final instructions. At this delay, however, the ship's company, who engaged in the voyage on proportionable shares of the profits resulting from it, began to express their dissatisfaction; while the owners of different properties in the South Seas, particularly Messrs. Enderby's, Messrs. Champion and Messrs. Methers, being anxious to convey early intelligence to their vessels, of the situation of our domestic affairs, and the change that appeared to be taking place in Europe, earnestly and anxiously recommended me to proceed on the voyage, without any other instructions than such as had been verbally communicated to me, and the maps with which Mr. Stephens had been so kind as to furnish me. It was, however, thought necessary that, previous to my departure, I should apply for the usual letter of leave of absence, which I immediately obtained.

After

After this long detention, and the great expences which accompanied it, I should not have been induced to undertake such a voyage, for the mere casual advantages which the fishery might produce, if I had not received the strongest assurances from the beginning, that, if I executed the commission assigned me by the Board of Admiralty, I should not fail of particular promotion; and that in any general promotion which might take place, during my absence, I should not be forgotten.

Though my former voyages were principally undertaken with the views of commercial advantage, I was never inattentive to the advancement of nautical science: my observation was always awake to every object which might instruct myself and enable me to instruct others; and I constantly committed my thoughts to paper as they arose in my mind from the appearance of things around me, or the circumstances, whatever they might be, in which I happened to be involved. I cannot be

supposed

fuppofed to poffefs any claim to literary qualifications, which are only to be obtained in the calm of life, while fo many of my years have been paft amid the winds and waves, in various climes and diftant oceans. Neverthelefs, I poffefs the hope that my work may contain fome portion of profeffional utility, or I fhould not have prefumed to prefent it to Mr. Stephens, as a teftimony, humble as it may be, of my gratitude for his condefcending kindnefs to me.

The owners, Meffrs. Enderby and Sons, being perfectly acquainted with the intention and bafis of the plan on which the voyage was to be conducted, made out their orders in a manner altogether comformable to the views of Government; and as a proof of their confidence, furnifhed me with full powers to act as I fhould think beft for our common intereft.

I accord-

I accordingly joined the ſhip on the 24th of December, 1792, but was unfortunately detained by contrary winds. During this accidental delay, a bounty of five pounds was offered for ſeamen to enter into his Majeſty's ſervice, which proved too great a temptation for all my crew to withſtand, as it afforded a favourable opportunity to thoſe who felt no compunction at robbing their employers of the advance given them to perform a long voyage, to ſhelter themſelves under the proclamation. I employed every poſſible precaution to prevent the defection of any of my people; neverthelefs, three of them contrived to make their eſcape: and I could not obtain any to ſupply their place, but by paying a bounty equal to that of the Royal Navy. On this condition I procured three men, natives of the Iſle of Wight.

The firſt day of January, 1793, at length arrived, and by this delay, we loſt the profpect of obtaining the premium given by the Britiſh

Govern-

Government to whalers failing round Cape Horn, without clearing out again, for which I received the owners inftructions: but the collector at Cowes informed me, that it could not be allowed.

On the fecond day of January, in the afternoon, the weather promifing to be fair, and the wind inclining to the North, I dropped down to Yarmouth Roads. We had a thick fog and Southerly wind all night, and during the next day and following night, there was an heavy gale to the South, with drizzling rain. At noon on the fourth, the wind fhifted to the Northward and brought clear weather, with which we failed.

CHAP.

CHAPTER I.

PASSAGE OF THE RATTLER FROM ENGLAND TO RIO JANEIRO.

THE flant of wind with which we fet fail continued to be fair no longer than midnight, and we were obliged to ply to windward until the feventh of January at noon; when, being clofe in with the Start, and threatened with a gale of wind from the Southward, we bore up and anchored in Torbay for the night. The wind at day-light fhifting to the North North Weft, we weighed and ftood out of the bay. This fpirt carried us as far as the Eddyftone Light-houfe, when it again fhifted to the Southward and blew frefh. We had now to beat off a lee-fhore, and, by the prefs of fail which we were obliged to carry, in order to accomplifh that object, and to get ten leagues to the Southward and Weftward of the Lizard, we had three feet water in the hold from a leak in

1793. the trunks that were funk to the hawfe for the cables, in confequence of a fpar-deck being fixed to join the fore-caftle and quarter-deck, and bringing the cables on the upper deck. At this time the gale had fo increafed, as to reduce our fails to the three ftorm ftay-fails; and, at times, during the twenty-four hours it lafted, we could not carry all of them, from the rain, hail, fnow and blowing weather which we had experienced more or lefs every day, fince we failed: while our crew confifted of no more than feventeen, officers and feamen, with three landmen and five boys, to work a fhip that, in his Majefty's fervice, had a complement of 130 men: and all the alteration I had made was cutting four feet off the lower yards, two feet off the top-fail yards, and reducing the fails in proportion. Thus weak handed, we were all obliged to be on deck when there was an extra duty, which our fituation and the ftate of the fhip often required. The crew alfo, at this time, began to grow weary, and, in addition to our other exertions, it was neceffary to keep the pump in continual employment. We were, however, greatly indebted to the marine barometer, as it warned me againft making fail when there was an appearance only of moderate, and to fhorten fail on the approach of foul and

January 13. tempeftuous, weather. At length, however, on the thirteenth at noon, when we were within a few leagues of the Lizard, the wind fhifting to the North Weft, and from thence to
the

the North Eaſt, a ſtrong wind and great ſwell carried us to Madeira in ſix days; on one of which alone we had fair weather. On the twentieth I had run that diſtance by two of Arnold's time-pieces and account: it alſo blew a ſtrong gale, very variable, with dark cloudy weather and heavy rain. I had not made any obſervation this day to be relied on, but ſuch as pointed out to me the Iſle was not far diſtant, and that it became abſolutely neceſſary for me to aſcertain our true ſituation before night. I depended on the qualities of the ſhip for clearing the land if caught on a lee-ſhore, and accordingly ſhortened ſail to cloſe-reefed main-top-ſail and fore-ſail. We then hove too and houſed our boats: but we had no ſooner bore up, than, half a mile on the lea-beam, we deſcried the Deſerters Rocks: and as it was impoſſible to weather them on the tack we were then on, we wore and ſtretched out between Porto Sancto and the Eaſt end of Madeira; while it blew ſo heavy at intervals, that the ſhip lurched three ſtreaks of the main-deck under water: at the ſame time, ſhe made a better way through the water than we could expect or would generally be believed. When the gale had ceaſed, calms, light winds, and baffling weather, prevented our clearing the Weſt end of Madeira, until the evening of the twenty-ſecond of January.

My preſent intention was to paſs in ſight to the Weſtward of the Canaries; and at noon, on the twenty-ninth

1793.

ninth inftant, had the weather not been fo dark and clouded, we ought, by our obfervations, to have feen fome of them. At four in the afternoon we got fight of the Ifland Ferro, diftant about five or fix leagues. From hence I was perfuaded to get in the Longitude of 21° Weft, in the Latitude of the Ifle of Sal one of the Cape de Verds, and run down in that parallel for the Ifle with an expectation of catching whale. This was not merely a curious inclination, but a fenfe of duty, which infpired the wifh to begin my acquaintance with that bufinefs, at as early a period of the voyage as poffible. Dark, hazy and cloudy weather accompanied us all the way from the Canaries, and our rigging was covered with duft of the colour of brown fand, as if it had been laying on fhore. We ran the diftance by watch and reckoning to a few miles, but the continuance of hazy weather prevented our feeing it: and as it blew ftrong with a heavy fea, it was the whaling mafter's opinion, with fuch weather we could do nothing with fifh, if we fell in with them: I accordingly bore up, and run down the lee-fide of Bonavifta. Light winds prevented our

Feb. 10. croffing the Equator until the tenth of February, at midnight, in Longitude 24° 30′ Weft of Greenwich, and all the fifh we had as yet caught, were a fhark and a porpoife.

In the Latitude of 19° South, and Longitude 25° Weft, we loft the South Eaft trade wind, which had accompanied us

from

from 4° North: when a wind from the North East fell in with us, and continued until we reached the Latitude 21° 30′, and Longitude 36° West. At this time we had changeable weather, with lightening heavy rain, and a swell from the South West. The wind then shifted to the Southward and Eastward, and blew in that direction until the twenty-third of February, when, at midnight, we made Cape Frio; but calms and light winds prevented our getting into Rio Janeiro, until the twenty-fourth. We there found Governor Phillips on his homeward bound passage from Botany Bay, in the Atlantic Transport; and, on the following day, a South Whaler came in to stop a leak.

The rigging of the Rattler was in such a state as to require considerable repairs: the new work also wanted caulking; and that no further time might be lost, the season being already far advanced for doubling Cape Horn, I hired natives to supply me with water, as well as a couple of caulkers to assist our carpenter in caulking the ship and stopping our leak, which was under the hawse-pieces. At length, all our repairs being compleated, and our stock of provisions, including two live bullocks, being got on board, by the fifth of March, we set sail with the Mediator, the whaler already mentioned. We parted, however, with her, on getting out of the harbour, and passed, between the Rasor and Round Isles, to the Sea.

Governor

1793.   Governor Phillips failed the day before me, and was fo kind, among other civilities, as to be the bearer of a table of the rate of the time-keepers to Phillip Stephens, Efq. as well as of letters to the owners and my friends. On my return from taking leave of him, and at the diftance of about two cables length of his fhip, I ftruck with an harpoon the largeft turtle I ever faw: it weighed at leaft five hundred weight. Governor Phillips, on obferving our exertions on the occafion, immediately fent a boat to our affiftance, and I requefted his acceptance of the prize.

On the arrival of a fhip off Santa-Cruz at the mouth of Rio Janeiro, the Patrimore or harbour mafter comes on board, takes charge of the veffel, carries her into the harbour and moors her in a good birth. Sometimes the mate is firft taken out, as was the ceremony with me, to undergo an examination; but the captain is not fuffered to leave the fhip without orders; neither will any fupplies be admitted until a vifit has been made by the officers of police, to enquire into the health of the crew, from whence you come, whither you are bound, what is the particular object of your voyage, and the time you wifh to ftay. The mates are then taken on fhore to be examined, when their declaration with that of the commanding officer, is laid before the Viceroy whofe official permiffion muft be given before any commercial
intercourfe

intercourfe can take place between the ſhip and the ſhore: the captain and officers muſt alſo ſign a declaration, acknowledging that they and their crew conſider themſelves as amenable to the laws of the country, while they remain in it.

The land-breeze, at this place, commences in the evening, and generally continues until the morning; the length of time it blows, as well as its ſtrength, depends on the different feaſons of the year; and all veſſels leaving this port muſt take the advantage of the land or leading wind, the mouth of the harbour being too narrow to turn out. We had no ſooner left the harbour, than the Patrimore took his leave of us. It may be proper, however, to remark in this place, that the commanders of merchant veſſels are required to give one day's notice, previous to their falling from this port.

CHAP.

CHAPTER II.

ATTEMPT TO DISCOVER ISLE GRAND; AND PASSAGE ROUND CAPE HORN.

1793.

March 17.

ON leaving Rio Janeiro, I ſtood away to the Southward and Eaſtward to ſearch for the Iſland Grand, which is ſaid to lay in the Latitude of 45° South, and was the firſt object recommended to me by the Board of Admiralty. On the ſeventeenth of March, at noon, Latitude by obſervation 39° 33′ South; Longitude by the mean of chronometers 34° 21′ Weſt; and by account, 34° 25′; variation, 9° Eaſt. The ſea appeared of a pale green, and we ſaw many birds, ſome of which were ſaid by ſeveral of my people, to be of thoſe kinds which are ſuppoſed to indicate the vicinity of land; ſuch as ſand-larks, and a large ſpecies of curlew; but neither of the latter kinds of birds were ſeen by me. From noon of the ſeventeenth, until ſix in the evening, the wind blew from the South, South Eaſt, and we ſtood away to the Southward and Weſtward: it then became calm

calm and continued so till midnight; when it blew from the North West, being, at this time, in the situation which I had often heard my old commander, Captain Cook, mention, as the position of the Isle of Grand: I accordingly entertained great expectations of seeing it; more especially as the birds appeared in great numbers during the whole of the day. In the evening we stood away to the Southward, in which direction I continued my course for the night. At day-light, on the eighteenth, the surface of the water was covered with feathers; and frequently in the forenoon we passed several birch twigs, as well as quantities of drift-wood and sea-weed. These appearances continued until noon of the same day, when our observation was in Latitude 40° 12′ South: Longitude by observation of Sun and Moon, 35° 34′ West; and by mean of chronometers 34° 8′. At this time the appearance of the sea had changed to a dirty green; which could not be the effect of the sky, as it was very clear: those tokens of land induced me to heave to and try for soundings with an hundred and fifty fathoms of line, but got no bottom, we had no sooner got the lead in, when to our great astonishment, at three or four miles distance from us, the whole horizon was covered with birds of the blue peterel kind. At the same time black whales were seen spouting in every direction, and the boats pursued one to harpoon

1793

March 18.

1793. harpoon it, but without succefs. Indeed, we were not very solicitous to kill black whales, and willingly gave up the chafe at this time, to make all the fail we could, and to exert our utmost efforts in order, if poffible, to difcover the land before night; which every one on board had poffeffed themfelves with the idea of feeing, although at fuch a confiderable diftance from the Latitude in which it was fuppofed to lay.

During this afternoon we paffed feveral fields of fpawn, which caufed the water to wear the appearance of barely covering the furface of a bank. At fun-fet we could fee as far as twelve or fourteen leagues; but did not perceive any other figns of land than the great flight of birds which continued to accompany us, and they were fo numerous at times, that had they all been on the wing together, and above us, inftead of rifing in alternate flocks and fkimming after the whales, the atmofphere muft have been altogether darkened by them. And the number of whales in fight prefented a fair opportunity of making a profitable voyage in the article of black oil; but my predominant object was to fulfil the particular fervices recommended to me by the Lords of the Admiralty; and in one point I had at this moment, the moft flattering hopes of fucceeding.

Towards

Towards the evening, the barometer fell, and the weather began to be cloudy: but I continued ſtanding to the Southward with a freſh breeze till midnight, when we hove to and founded; but did not find ground, with one hundred and ſeventy fathoms of line. The gale was increaſing every hour with an heavy ſea; and, by day-light, we could only carry cloſe-reefed top-ſails and fore-ſail. The weather was dark and hazy, the ſea aſſumed a deep lead-colour, many birds and whales remained with us, and we paſſed large quantities of ſea-weed. At noon we were in the Latitude of 43° 3′ South, and Longitude 35° 38′ Weſt. Here we founded, but found no bottom: neverthelefs, every circumſtance ſtrengthened our conjectures that we were nearing the land, which induced me to proceed on my courſe, although it continued to blow hard from Weſt North Weſt. At midnight we hove to, and founded with one hundred and ſeventy fathoms of line, but found no bottom. At day-light we founded again with two hundred fathoms of line, and were equally unſuccefsful. We now made ſail, and at noon our Latitude was 44° 51′; Longitude by obſervation, 34° 59′; and by mean of chronometers 33° 53′ 30″ Weſt.

The birds leſſened greatly in numbers, and with them our hopes of finding the land which was the object of our ſearch. I continued, however, to cruize about for ſeveral

1793.

successive days near this Longitude, but saw nothing to encourage any further endeavours.

The season was now far advanced for doubling Cape Horn and it appeared to me, that the most rational course I could take, would be to run down West to the main land of Patagonia, in the Latitude in which the Isle of Grand is placed; as we were now to the Eastward of Mr. Dalrymple's position of it*: so that if it was not found in that Latitude, I might,

---

Extract from Mr. DALRYMPLE.

* In the Latitude of 45° South, there is a very large, pleasant island, discovered by Ant. La Roche, a native of England, in his passage from the South Seas, in the year 1675. The Spanish author who gives the abstract of La Roche's voyage, printed according to him, in 1678, says, " That La Roche, leaving the land, (discovered by him in 55° South, and which was since seen by the Leon, in 1756,) and sailing one whole day to the North West, the wind came so violently at South, that he stood North for three days more, till they were got into 46° South, when thinking themselves then secure, they relate, that directing their course for the Bahia de Todos Santos, in Brazil, they found, in 45° South, a very large, pleasant island, with a good port towards the Eastern part; in which they found wood, water and fish, they saw no people, notwithstanding they stayed there six days." The size of this island is not mentioned in the Spanish abstract; but the expression, Muy Grand, very large, and the expectation of finding inhabitants, seemed to indicate that it is of great extent.

The existence of this island, and, in some measure its extent, is confirmed by other authorities: for Halley, near this Longitude, in about 43° South says, " the colour of the sea was changed to pale green, and in 45° South he saw abundance of small sea-fowl and beds of weeds." Funnel, in his passage, into

the

might, on my return, fearch for it in the Latitudes of 40° and 41°, having ftrong reafon to believe, that there is land in or near thofe Latitudes, but to the Eaftward of the Longitude which I croffed; as otherwife, I am at a lofs to account for fuch a quantity of birch twigs, fea-weed, drift-wood and birds as were feen in that fituation. Some of thefe birds appeared to be quite young, from the difficulty with which they feemed to ufe their wings; though that circumftance, it is poffible, might have proceeded from their being gorged with fea blubber, with which the furface of the water was covered.

From the land difcovered by Monfieur La Roche, in Latitude 55° South, which I touched at with Captain Cook, in the year 1771, who named it Georgia, I am difpofed to believe, that the Ifle of Grand alfo exifts, and that my not being able to find it, arofe from an error in copying the Latitude

---

the South Sea, alfo mentions figns of land from about 40° South, near this Longitude. The Naffau fleet, 1624, had alfo figns of land here, fo as to think themfelves near the Southern continent.

Thefe teftimonies and the Leon, in 1756, finding the other land mentioned by La Roche, leave very little reafon to doubt his veracity: and, if there is fuch an ifland, fituated in the middle of the ocean, in 45° South Latitude, it cannot fail of being a very temperate and pleafant country, in a fituation very favourable for carrying on the whale fifhery and others, and alfo for the profecution of any commerce, which may be found in the countries to the South.

tude given by La Roche: nor can I doubt, from the quantity of whales I perceived near its fuppofed fituation, that it would prove a much greater acquifition than the Ifland Georgia, to which many profitable voyages had been made for feal fkins alone.

This route, however, will be of fome advantage to Britifh navigators; even if no land fhould be difcovered according to our expectations, as it will tend to undeceive the mafters and owners of whalers, who have entertained an opinion that the black whale was never to be found in bodies, fo far to the Eaftward : for, if half the whalers belonging to London had been with me, they might have filled their veffels with oil.

March 23. The autumnal equinoctial gale came on us the twenty-third of March, and held upwards of four days, with frequent claps of thunder, accompanied by lightening, hail and rain. It blew as hard as I ever remember, and, for feveral hours, we could not venture to fhew any fail. At the fame time a whirlwind or typhoon arofe to windward, from whence in one of the fqualls, two balls of fire, about the fize of cricket balls, fell on board. One of them ftruck the anchor which was houfed on the fore-caftle, and burfting into particles, ftruck the chief mate and one of the feamen, who

fell

fell down in excruciating tortures. On examining them several holes appeared to have been burned in their cloaths which were of flannel: and in various parts of their bodies there were small wounds, as if made with an hot iron of the size of a sixpenny piece. I immediately ordered some of the crew to perform the operation of the Otaheiteans, called Roro mee *, which caused a considerable abatement of their pains, but several days elapsed before they were perfectly recovered. The other ball struck the funnel of the caboose, made an explosion equal to that of a swivel gun, and burned several holes in the mizen-stay-sail and main-sail which were handed. At the height of it the barometer was 28°. The alarm which we may be supposed to have experienced during the whirlwind, was not allayed by the noise of the birds, who, not considering the ship to be a place of safety, as is the case in common gales, appeared, by the violence of their shrieks and the irregularity of their flight, to be sensible of the danger: for as the squall approached them numbers plunged into the sea, to avoid it; while those who could not escape its influence, were whirled in a spiral manner out of sight in an instant. It very fortunately reached us only within two cables length of each beam, and so passed a-head of the ship to the North. From our first seeing, to our losing sight

1793

---

* Roro mee. It consists in grasping the fleshy parts of the body, legs, and arms, and working it with the fingers.

1793. sight of it, was about half an hour. In this gale, I lost the greatest part of my live stock, together with all the vegetables that hung at the stern of the ship.

April 8. On the eighth day of April, in Latitude 50°, we struck soundings off the North West end of Falkland's Islands. The whole way I saw plenty of black whales; and two days before striking soundings, we perceived a shoal of spermaceti whales, apparently bound round Cape Horn; but our boats being all housed and well secured for doubling the Cape, we did not pursue them. In this course I ran directly over the situations in which the Isle of Grand is placed in all the charts, without discovering any appearance of land. On the

9. ninth, in the afternoon, we struck soundings in sixty-five fathoms off the West end of Falkland's Islands; but the violence of the wind and the thick weather prevented me from making an accurate observation. The shallowness of the bottom induced me to believe, that I was not so far to the West as the watch gave me. When by my calculation I was to the South of Falkland's Isles, I stood away for Cape Saint John, Staten Land. The winds were variable North West, South, South East, East, and North East. The greatest depth at which I found bottom, was ninety fathoms; and then no bottom at one hundred and fifty fathoms.

On

On the eleventh at midnight when I fuppofed myfelf off Cape Saint John, we founded and ftruck ground at ninety fathoms, fmall dark ftones. In this depth of water I did not confider myfelf as far to the Eaftward of the Cape as I wifhed, for which reafon, I hauled on a wind, and beat to windward, with the wind at North Eaft and Eaft North Eaft, till feven o'clock in the morning, when having no foundings at one hundred and fifty fathoms, I bore up for Cape Horn. On the twelfth at noon, the wind drew round to the South, South Eaft, as the preceding night indicated, by the cold being fo fevere, that fome of the crew were froft bitten for feveral hours, and the fhip and rigging covered with fnow and ice. After this, the wind inclined to the Weft of South, which was foon fucceeded by moderate weather and fmooth water; this was alfo of fhort duration, for it changed gradually round, until it got to the Eaft, and at midnight on the thirteenth, it fhifted fuddenly in a fquall of rain to the South, and brought me on a lee-fhore.

At day-light we faw the Ifles of Diego Ramieres, bearing North by Eaft, three or four leagues; and I make them by obfervations corrected, in Longitude 68° 58′ Weft; and in Latitude 56° 30′ South. They appeared to lay in an Eaft and Weft direction. The Weftern Ifle, which is the higheft, is furrounded with fmall iflets; and the circumference of the whole

whole may be nine or ten miles. With the affiftance of telefcopes, it appeared to be entirely barren, though it may be an afylum for feals; there being many about us at this time, as well as white crows. Thefe birds refemble in fize and figure, the dun crow, which I have frequently feen in Hampfhire, in the winter feafon, and is probably a bird of paffage. We daily faw thefe crows, from the Latitude of the Falkland Iflands, until we had doubled Cape Horn. They all appeared to come from the Eaftward and Southward; perhaps from Sandwich land, and to be bound to the main land of America. Several of them were caught, but could not be preferved alive. It may be remarked, that I never faw any of thefe birds at Cape Horn in my former voyages.

CHAP.

CHAPTER III.

REMARKS ON THE NAVIGATION ROUND CAPE HORN.

I HAVE doubled Cape Horn in different feafons; but were I to make another voyage to this part of the globe, and could command my time, I would moft certainly prefer the beginning of winter, or even winter itfelf, with moon-light nights; for, in that feafon, the winds begin to vary to the Eaftward; as I found them, and as Captain, now Admiral, Macbride, obferved at the Falkland Ifles. Another error, which, in my opinion, the commanders of veffels bound round Cape Horn commit, is, by keeping between the Falkland Ifles and the main, and through the Straits Le Maire; which not only lengthens the diftance, but fubjects them to an heavy, irregular fea, occafioned by the rapidity of the current and tides in that channel, which may be avoided, by paffing to the Eaftward. At the fame time, I would recommend them to keep near the coaft of Staten Land, and Terra del Fuego, becaufe the winds are more variable, in with the fhore, than at a long offing.

If it fhould be obferved, that a want of wood and water may render it neceffary for veffels to ftop in the Straits Le Maire, I fhall anfwer, that there is plenty of water at the Falkland Ifles; and Staten Ifland, not only abounds in both,

1793. but possesses several excellent harbours. I first visited this place with Captain Cook, in the year 1774; and, on my outward-bound passage to the North West coast of America, in the year 1786, as commander of the merchant ship, Prince of Wales*, I wooded and watered there, and left a party to kill seals. For my own part, I do not perceive the necessity, according to the opinion of different navigators, of going to 60° South. I never would myself exceed 57° 30', to give the Isle of Diego Ramieres a good birth, or, if winds and weather would permit, make it, for a fresh departure, had I not taken one at Cape Saint John, Staten Land, or the East end of Falkland Isles. Staten Land is well situated as a place of rendezvous both for men of war and merchant ships; while the harbours on the North and South sides, which are divided by a small neck, would answer the purpose of ships bound out, or home. But the North side offers the best place for an establishment, if it should ever be in the view of our government to form one there†.

CHAP.

---

* To the owner of this ship I was first introduced by one of the most eminent merchants of the City of London.

† If the navigation round Cape Horn should ever become common, such a place we must possess; and agreeable to the last convention with Spain, we are entitled to keep possession of it, and apply it to any purpose of peace or war. Great advantages might arise from such a settlement, from whence the black whale fisheries might be carried on to the South Pole, in the opinion of all the North Greenland fishermen, with whom, I have conversed on the subject. Besides, it is one of the easiest land-falls a sailor can make. In order to render this place a defensible, and protecting settlement, many experienced men, lieutenants, in his Majesty's navy, might be found, at very little extra expence to government, to live in a situation, which would be far preferable to many stations in Norway, that I have seen. The officer placed there, should be invested with full powers to regulate all fishers, fishing in those parts, or navigating round Cape Horn, that stop at the port.

CHAPTER IV.

ROUTE FROM CAPE HORN TO MAKING THE COAST OF CHILI, AND THE ISLES ST. FELIX AND ST. AMBROSE.

WE doubled Cape Horn on the 11th of April, the day three months on which we departed from England, after having stopped at Rio Janeiro, during the space of ten days; and proceeding from thence, four hundred leagues to the Eastward, in search of the Island of Grand. On making Diego Ramieres Isles, we stretched well into the Westward of them, with the wind at South, South West; and, at midnight, tacked to the Southward and Eastward. During the following five days, we had the wind from West, and South West, and mostly with an heavy gale, and a tremendous sea, and as bad weather as could be produced by hail, rain, frost and snow, which drove us as far to the Southward as 59°, and back to the East as 69°. We neverthelefs strove well against it, and the crew being in good health.

1793.
April 11.

1793. health and spirits, we made sail, if it became moderate, only for half an hour; and, at length, fairly beat round the land of Terra del Fuego. No sooner had we attained this object, than we had fine weather, with a Southerly wind; which promised favourably, to my present intention of examining Wager Island, on the West coast of Patagonia, which we saw on

April 28. the twenty-eighth, at midnight.

Wager Isle is high and rugged, and may be seen at the distance of fourteen or fifteen leagues. It is about five or six leagues in length, and lays, by compass, nearly in a North and South direction, with many islets off both North and South ends. I place the body of it in Latitude 46° 30', and Longitude 76° West. On the western side, where nothing grows but a small quantity of green moss, it wears a very barren appearance, and the distant hills, bearing East 25° North, I believe, were mountains on the main land, covered with snow. Capt. Cheap, who commanded the Wager, one of Lord Anson's squadron, has given a full description of this island, where he was unfortunately cast away*. My design in making it, was to obtain some knowledge of Anna Pink Bay and Harbour, but the coast was so forbidding, and the weather of such a dark, hazy, and wintry aspect, as to discourage me from persevering in it. Besides, having doubled Cape Horn at the precise time of the year when Lord Anson went round it, and being at Wager Isle, within a fortnight of the time, when

Captain

---

* In the year 1741.

Captain Cheap was shipwrecked there, I was discouraged from paying any further attention to it. The inlet, which was the object of my search, is not a mile wide; a space, which can be descried, but on a very near approach. The Anna Pink did not see it, until she was within a mile or two of the rocks and breakers, among which it lies; and although they may shew themselves, the depth of water is so great in the bay, that when found, no whaler will attempt to make it, because he cannot trust to his anchors. I tried for sounding several times off Wager Isle, but got no bottom; neither was the colour of the water so much changed here, as the day before we made the land.

By the Anna Pinks supposed Latitude of that place, and my own observations, I have no doubt, as was conjectured, at the time, that the crew of the Wager heard the Anna Pinks guns; and that she lay under the main to the East of Wager Island*. If the design proposed by Captain Cheap had been adopted, of coasting in the boats, it is more than probable that it would have succeeded; and the well-known distresses of that officer and his crew would then have been avoided. The many escapes and voyages which, from shipwreck, views of gain, and other causes, have been made and performed in boats within these few years are

---

* The Anna Pink was a victualler belonging to Lord Anson's squadron, and driven into this port in distress.

are, I presume, pretty generally known. My long-boat, which was not more than twenty-eight feet in the keel, coasted it in the depth of winter and in a tempestuous clime, from 38° North to 50° North; and part of the same course back to rejoin me. The greatest part of the cargoes of ships voyaging to the North West Coast of America, have been collected in boats. The long-boat of the ships I commanded in my first voyage thither, coasted from 60° North, many leagues to the Southward, as well as in various bays and inlets which the ship could not enter; whereby a more particular knowledge of that country has been obtained, as will appear on the publication of Capt. Vancouver's voyage. The known spirit of enterprise and speculation, possessed by the British merchants, and which is not equalled, by those of any other nation, will again shew itself, when a peace takes place; and the inland countries in the Southern parts of America, including the East and West coasts of Patagonia, Straits of Magellan, and Terra Del Fuego, will, I doubt not, be traversed and explored, from the information of Mr. Falkner, who resided near forty years in that country, and published an account of his travels in 1774. He remarks, that the Eastern coasts abound with ostriches, whose feathers are known to be so valuable, besides otters, hares, rabbits, and other animals, yielding fur. He likewise mentions the articles of trade necessary for the Indians, and describes the Western side as abounding also with ostriches, as well as hares of an extraordinary size,

and

and black rabbits; whose fur is superior to that of the beaver. The otters and hares I have seen, and their skins would be a valuable article of commerce in China.

He further declares, as his decided opinion, that it would prove very disadvantageous to the Spaniards, if the English were to form any settlement to the South of Brazil; there being several rivers, which communicate with the Western side of America; and he gives a particular description of the bay St. Fondo, and river Colorado.

This idea, however, had not awakened the attention of any court, till disputes arose about Nootka Sound, in which I unfortunately, had so great a share*. I cannot pass over in silence the very

---

* Mr. Thomas Falkner was the son of a surgeon of eminence at Manchester, and was brought up in his father's profession, for which he always manifested the most promising dispositions. To complete his professional studies, he was sent to London to attend Saint Thomas's Hospital; and, happening to lodge in Tooley-street, on the Surry bank of the Thames, he made an acquaintance with the master of a ship, employed in the Guinea trade, who persuaded the young surgeon to accompany him in his next voyage in his professional capacity. On his return to England, he engaged to go in the same situation on board a merchant ship to Cadiz, from whence he continued his voyage to Buenos Ayres, a Spanish settlement on the River La Plata. Here he fell sick, and was in so dangerous a state when his ship was ready to depart, as not to be in a condition to be carried on board: so she sailed without him. The Jesuits, of which there was a college at Buenos Ayres, nursed him during his illness with the greatest care and kindest assiduity;

1793.

very fingular circumftance, that while the Spaniards were negotiating with Great-Britain, for arranging the difference between them, in an amicable manner, they actually fettled every port in the Atlantic, as far South as the Straits of Magellan;

---

affiduity; and perceiving the very great advantage which they would derive, in their miffions, from poffeffing a brother who was fo well fkilled in medicine and furgery, fpared no pains to win his affection and fecure his confidence. In fhort, they fo worked upon his mind, as to perfuade him to enter into their college and finally to become one of their order. He now entered upon his miniftry among the Indians, who inhabit the vaft track of country between the River La Plata and the Straits of Magellan. His fkill in the cure of difeafes, and in performing chirurgical operations, together with his knowledge of mechanics, rendered his miffion fuceefsful beyond example. In this country he remained near forty years, and was among the perfons appointed by the Spanifh Government, to make a furvey of the coafts between the Brazils and the Terra del Fuego, Falkland's Iflands, &c. When the fociety of Jefuits was diffolved, he was fent back to Spain, and after an abfence of near forty years, arrived in his native country. Soon after his return to England, he became domeftic chaplain to Robert Berkeley, Efq. of Spetchley, near Worcefter, a Roman Catholic gentleman of diftinguifhed knowledge, moft refpectable character, and large fortune. There he wrote the account of Patagonia, which has been quoted in this volume, and was afterwards publifhed with a map corrected from that of D'Anville, according to his own obfervations. Mr Falkner poffeffed a very acute mind, a general knowledge and moft retentive memory. Of his medical experience and practice, I have heard phyficians of eminence fpeak in the higheft terms of commendation. His manners, as may be fuppofed, from the tenor of his life, were at once fingular and inoffenfive: and he retained fomewhat of his Indian habits to the laft. He died, as I have been informed, about the year 1781.

gellan; and attempted it at New Year's Harbour, Staten Land. At that place I left a factory in my first voyage, in 1786; but the ship which was sent to carry them supplies being cast away, they quitted it in boats. Another body of English merchants, engaged in a similar establishment, and was there, when the Spaniards arrived, which induced the latter to abandon their design, and, by a violent gale from the Eastward on the night after their arrival, their ship was either wrecked or driven to sea between the New Year's Isles.

1793.

This was not the only political manœuvre of the Spaniards; for they intimated to Lord Saint Helens, as his Lordship informed me, that their settlements in the Californias, extended a long way to the North of Port St. Francisco. This circumstance, I represented to his Lordship to be altogether untrue, as my long-boat had coasted from Saint Francisco to Nootka, and saw no settlements. I have no doubt, nevertheless, of its being their design to settle the intermediate space, as well as the Sandwich Isles, for which they had made every preparation, at the time I was released from their naval port of Saint Blas, in the Gulph of California. Indeed, while I was on my present voyage, they settled the Port of Sir Francis Drake, where I wooded and watered, in my preceding voyage. But it is probable that Captain Vancouver, whose voyages are now preparing for publication, will give a more particular account of this settlement.

1793.

If such an enterprize has not commenced, the period, however, cannot be distant, when a commercial search after furs, seals and whales, will lead adventurers to traverse the Coasts of Terra del Fuego and Patagonia, for them. The whales and seals are grown shy, and become scarce in other parts where they have been hitherto taken, so that new haunts must be resorted to, in order to keep up the spirit of our fisheries; and those coasts will probably employ fifty or sixty sail of merchant ships, as they abound not only with black whales and seals, but the sea elephant, and the sea lion.

The Southernmost settlement of the Spaniards, known on the Western side, is Brewers, or English Harbour, in Latitude 44° 20′ South*. To the South of it, is a great archipelago, and many deep inlets, which perhaps, are unknown to the Spaniards.

As from the circumstances already mentioned, I had given up all search after Anna Pink Bay, I was, from the same cause, induced to relinquish my design of looking into Brewers Harbour, and did not make land again, till we were in the Latitude 38° 39′, when we saw the Isle Mocha.

May 1.

In 40° South, we saw spermaceti whales, but did not take any till the first of May, when we made the Isle, the sea being then

---

* This Harbour is named from Mr. Henry Brewer who commanded a squadron of Dutch ships in 1642, belonging to the Dutch West-India Company on an expedition to the coast of Chili, who found in this port refreshments of every kind, and also secure and good anchorage.

then covered with them: but of six which we killed, and of four secured along side, the weather proved so bad, that we could only save the bodies of two fish, and one head.

We beat to windward off Mocha for several days, during which time we saw a great number of whales, but killed no more than two, of which we saved one. Several of my people getting hurt in flinching them*, and others breaking out in boils from the bad state of the weather, I was obliged to pursue my voyage; but, by ordering the sick two oranges every day each person, with a large glass of lime juice and water every morning, they soon recovered, as well as those also who were hurt in whaling.

I place Mocha Isle in Latitude 38° 24′, and Longitude 75° 00′. It is of an height to be seen at the distance of fifteen or sixteen leagues, in clear weather, and on approaching it, its summit appears rugged. It is about three or four leagues in length, laying in a North and South direction by compass. The Northern part of the island descends gradually into a low, sandy point, or long tongue of land, on which is a rock or cross, that has the semblance of a sail. The South point, at the distance we were, appeared to end in a more abrupt manner, and there extends

---

* This expression is used for taking off the fat from the fish.

1793. extends from it, for a confiderable diftance, a range of fmall rugged rocks, fome of them on a level with the water; while others rofe boldly from it, fo that it was a matter of fome doubt with me, whether they compofed an actual part of the ifland. Breakers alfo run off from it a great way to the Weftward, at leaft three leagues. There is a bare, white fpot on one part of the ifland, having the appearance, at an offing, of eight or nine miles, as if not belonging to it. From the great number of feals, which I faw off this Ifland, I fhould fuppofe that it muft abound with them. The main land of Chili, within the ifle, is of a moderate height, and as it appeared to me, about fix or feven leagues diftant.

The only account I have been able to obtain of this Ifland is from Mr. Wafer's voyage, who was there, with Capt. Davis, in the year 1686, at which time, they lay there during the fpace of eight days. He relates, that they wooded, watered and ftored themfelves with frefh provifions, particularly the guanaco*. In fhort, he gives a very interefting account of the neceffaries

---

Extract from Mr. WAFER.

* The ifland afforded both water and frefh provifion for our men. The land is very low and flat, and upon the fea coaft fandy; but the middle ground is good mould, and produces maize, wheat and barley, with variety of fruits, &c. Here were feveral houfes, belonging to the Spanifh Indians, which were very well ftored with dunghill fowl. They have here alfo feveral horfes: but that which is moft worthy of note, is a fort of fheep they have, which the inhabitants call

neceffaries which they procured. He fays the land is low and flat; but he muft certainly fpeak only of the Eaft fide, or North and South points: If whales are as plentiful in the offing in the fine feafon, as at the time when I was there,

it

---

call *Cornera de Terra*. This creature is about four feet and an half high at the back, and a very ftately beaft. Thefe fheep are fo tame, that we frequently ufed to bridle one of them, upon whofe back two of the luftieft men would ride at once round the ifland, to drive the reft to the fold. His ordinary pace is either an amble or a good hand-gallop; nor does he care for going any other pace, during the time his rider is upon his back. His mouth is like that of a hare; and the hair-lip above opens as well as the main-lips, when he bites the grafs, which he does very near. His head is much like an antelope, but they had no horns when we were there; yet we found very large horns, much twifted, in the form of a fnail-fhell, which we fuppofed they had fhed; there laid many of them fcattered upon the fandy bays. His ears refemble thofe of an afs, his neck fmall, and refembling a camels. He carries his head bending, and very ftately, like a fwan; is full chefted like a horfe, and has his loins much like a well fhaped greyhound. His buttocks refemble thofe of a full grown deer, and he has much fuch a tail. He is cloven-footed like a fheep, but, on the infide of each foot has a large claw, bigger than ones finger, but fharp, and refembling thofe of an eagle. Thefe claws ftand about two inches above the divifion of the hoof; and they ferve him in climbing rocks, holding faft, by whatever they bear againft. His flefh eats as like mutton as can be: he bears wool of twelve or fourteen inches long upon the belly; but it is fhorter on the back, fhaggy, and a little inclining to a curl. It is an innocent and very ferviceable beaft, fit for any drudgery. Of thefe we killed forty-three; out of the maw of one of which I took thirteen Bezoar ftones, of which fome were ragged, and of feveral forms; fome long, refembling coral; fome round, and fome oval; but all green when taken out of the maw: yet by long keeping they turned of an afh colour.

1793 it is an excellent whaling ground; and the Isle itself very convenient for the purposes of refreshment. For although Mr. Wafer, on his return, found that the Spaniards had destroyed all the produce of the Island, of which they could possess themselves, to prevent its affording relief to the Buccaneers; so long a period has since elapsed, that it is no wild conjecture to suppose, it may now have regained its former plenty.

We kept the land of Chili in sight, from the mast-head or deck, until we reached the Latitude of 26° 20′; in which parallel, the Islands of Saint Felix and Saint Ambrose, were supposed to lie, but at the distance of one hundred and fifty leagues from the main. When I first fitted out, my intention was to visit the Isles Juan Fernandes, and Masa Fuero: but, before I left England, Europe was in such an unsettled state, as to induce me to consider a rupture between Great Britain and Spain, as no improbable event, when the cruizers, of the latter nation, would certainly be stationed off these islands: besides, I had every reason to believe, that, at the Saint Felix and Saint Ambrose Isles, I should find refreshments equal to, those which Masa Fuero is known to afford. Besides, having learned, at Rio Janeiro, that Lieutenant Moss, of the royal navy, whose nautical abilities are known and approved, had been lately sealing at Masa Fuero, and taken great pains to ascertain their situation, I conceived it unnecessary, for me to go there for that purpose only.

All

All the way to the Latitude of Saint Felix and Saint Ambrofe, and alfo running down the Longitude to thofe ifles, we never faw a fingle fpermaceti whale, except a flinched one, the day before we quitted fight of the main; but there were a great number of feals along the whole coaft.

On the twentieth day of May, at day-light, we faw one of the Ifles of Saint Felix and Saint Ambrofe; and foon after the other. By four in the afternoon, we were within fix or feven miles of the Eafternmoft; when, accompanied by the whaling mafter, I made an attempt to land, as well as to find an anchoring ground. The ifle proved to be a rugged, perpendicular, barren rock, fixty or feventy fathoms in height; and, in its craggy breaks and fhelvings, feals had found a refting place. There was, indeed, an appearance of verdure on its fummit, which induced me to conjecture, that it is, by fome means or other, fupplied with moifture. But night coming on, and it beginning to fniffle and rain, with the wind far to the North, and no place to fhelter the boat, or where we could land, on the North fide, we returned on board. It was a fqually night, with fhowers of rain; and, on the twenty-firft, at day-light, as much wind blew from Weft North Weft, as would admit of double reefed top-fails, with very heavy fhowers, which continued the whole of the forenoon; fo that we were obliged to pafs to the Southward of the Eafternmoft ifle, which prefents the fame inhofpitable afpect as that of the North fide. At noon, the rain ceafed, and the wind became variable with cloudy

1793. weather and much sea. By sun-set, we got well up with the Western isle, and being moon-light, I sent the chief mate, in one of the boats, to fish during the night, as well as to take soundings round the isle for the best anchoring place; and in the morning to make an attempt to land. At noon, on the following day, he returned with plenty of fish of the species of cod and bream, weighing from four to six pounds each; and informed me he had taken soundings round the isle, and that the only bay was on the South side; but that he could not find any bottom except close to the shore, which was at eighteen fathoms depth and rocky: That he had also sounded, on the North side, round the isle, to eight fathoms, within half a mile of the shore, and found a sandy bottom; but beyond that, could find no ground, at thirty fathoms; and, that the late gale had occasioned so great a surf as to render it impossible to land. He added, that the island appeared to be covered with seals. I had by this time surrounded the isle, with the ship, and frequently tried for soundings, but no bottom could be found, with one hundred and seventy fathoms of line, at the distance of from four to six miles from the shore.

The whaling master and second mate attempted to land in the evening, the swell having considerably abated; but they could not accomplish their design. They set out again, at four

May 22. o'clock the next morning, twenty-second, with a similar design; and,

and, having with great rifk and difficulty effected a landing, they traverfed the ifle, which produced nothing but a plant, refembling the common nettle, of a falt tafte and difagreeable odor. They could find no frefh water, and the foil was mere fand, from one to fix inches deep on a folid rock, and wafhed into furrows, as it appeared, by heavy rains. No land-bird, quadruped, or even infect, except flies, were feen on the ifland; but great numbers of birds-nefts, containing addled eggs: nor was there any kind of fhell-fifh. On the ifland, were the remains of feals and a quantity of decayed feal-fkins, fuppofed to have been left there by Mr. Ellis, (who vifited this place in the year 1791,) and probably, from the difficulty of taking them away.

Of the danger, of getting from this ifland, we had a very melancholy experience; as our people were upfet feveral times, before they got from the furf, and one of our beft feamen, was unfortunately killed, having his back broken, by the jolly-boat falling, upon him.

The only landing place, is on a fandy beach on the North fide of the ifle; and the tide ebbed on it, while they were on fhore, between fix and feven feet, and the ebb and flood runs to the Northward and Eaftward. At the time they landed, which was at fix in the morning, it was near high

high water, and when they got off, at two o'clock, P. M., it was low water. Neither, of thefe ifles is more, than five or fix miles in circumference, and they are diftant from each other four leagues and an half. The Eafternmoft ifle, appearing to be inacceffible; can never be of much ufe, except as a place for catching fifh or taking feals: But the other ifle, might be made to anfwer, as a place of rendezvous in war or peace. It contains a fpace, where tents might be pitched, and the fick accommodated, if the want of wood, water and vegetables, could by any means, be furmounted. As to the firft, an hull or two, of a prize, would afford a fufficient fupply; and as for the fecond, a ftill, might be provided, to diftil falt-water; and a fmall quantity of foil, would be fufficient, to raife fallad herbs.

A whaler, indeed, never wants wood or water; of the latter, fhe has always a very large quantity; becaufe, all her empty cafks, which fhe carries out for oil, are filled with water, by way of ballaft; and as it is to be hoped, that the fifheries will not be difcontinued, fhe might always leave her furplus quantity here, by ftarting it into a tank. A few buts of earth, might alfo be brought, and feveral kinds of fmall herbs, be raifed. Should this retreat be difcovered, by the capture of any veffel, it is fo fituated, that a fmall force would be enabled to defend it againft a large one. In the winter feafon, however, I cannot fuppofe, that any cruizer,

as

as the prevailing winds are Northerly and tempestuous, will attempt to anchor on a lee-shore; but, during the summer, when the winds are changeable, moderate and Southerly, I have no doubt, but ships might remain here in perfect security.

When South of the Western isle, the whole has the appearance of a double-headed shot; but the Eastern hummock is separated from it, by a very narrow reef, which divides it, as it were, into two isles; the lowest land, commencing from the reef, and joining the hummock to the West. There is also a remarkable small rock off the North West end, which, in most points of view, shews itself, like a ship under sail. These isles I place by observation corrected of Sun, Moon, Stars, and chronometer, between the Latitude of 26° 19′, and 26° 13′; and Longitude, 79° 4′, and 79° 26′ West.

CHAP.

CHAPTER V.

ROUTE OF THE RATTLER FROM THE ISLES SAINT FELIX AND SAINT AMBROSE, TO THE COAST OF PERU.

1793. Mr. Dalrymple conjectures, that, to the West of the Isles Saint Felix and Saint Ambrose, there are others, of the same name, which were called Saint Felix and Ambrose Rocks. To ascertain the truth of this opinion, I ran as far to the Westward, as 84°, when seeing neither land, birds or seals, to justify a belief that any such isles existed near this situation, I hauled on a wind for the coast of Peru, intending to make it, in Latitude 17° South; but, the wind hanging to the Eastward, I did not get on the fishing ground, until I was in the Latitude of 15° 30′.

There was now so large an extent of coast, in every part of which, I might meet with those British ships, employed in

spermaceti

spermaceti whaling, to whom, I was instructed, to communicate, the circumstances and situation of Europe, when I quitted it, that I did not think it necessary to beat again to the Southward. I was indeed, persuaded, that the greatest body of fishermen was to the Northward; as they would find the best season there, and might then return with the Sun, without being liable to the uncertainty of getting fish to the Southward, in the winter season; from whence, if they were not successful, they would be obliged to proceed to the Northward, and thus have two winters to encounter.

On the third of June, P. M. we were, within six or seven leagues of the Isle Lobas or Seals, near the port of Pisco, where we saw great numbers of that animal, and we had even fallen in with them, as far as fifty or sixty leagues from the land. Mr. Fresier says, that this isle is only one league and an half from the main land; but to me, it appeared to be twice that distance. He also adds, that the channel, between it, and the main land, is dangerous; but that, to the Northward of the isles, there is a smooth flat bank of sand, which forms a creek, where the sea is so still, that a ship can anchor there, in eight fathoms water, and might careen in safety. This island is of a moderate height, but, from the best observations I could make, in hazy weather, its coast appeared to be altogether barren to the Westward. This place offers a good port for whalers, or in time

1793.

time of war, for veffels of fmall force, to lay at, in order to watch an enemy; the land, being of fufficient height, to afford fecurity and concealment.

I continued my route along the coaft to the Northward, running under an eafy fail, or ftanding off and on in the day, and laying to at night. I never diftanced the land, more than fourteen or fifteen leagues, and was fometimes as near it, as two or three leagues. I cruized feveral days off Lima, at a fmall diftance from the Ifle Saint Lawrence, which forms the road of Callo* I kept near this fituation, in hopes of meeting fome veffel, which might afford me information, whether any Englifh fifhermen were in the road, and without any apprehenfion, of being known by the Spaniards, as the fuperior failing of my fhip, always left it to my own option, to fpeak with whom I pleafed.

June 6.

On the fixth of June, at fun-fet, I faw the dangerous rocks and fhoals of Ormigas, appearing like a fail, and laying nearly Eaft and Weft of Ifle Saint Lawrence. At noon our Latitude obferved was 11° 48′, the Ifle Saint Lawrence Eaft, 80° North, and the rocks of Ormigas, North 28° Weft, at the diftance of feven or eight miles. Thefe rocks are very dangerous; the loftieft part being little higher than the hull of a fmall fhip; and the fea breaks, for feveral leagues, around, and off, them. They are quite barren, and I obferved with my glafs, two croffes erected on them, which in a fhort time difap-

---

* In 1624, the Dutch fortified themfelves at this ifle, when they were making preparations to attack Lima.

peared. I concluded, therefore, that they were placed by fishermen, who are said to resort here from Lima, as signals, to engage in some kind of contraband trade: but I had taken the necessary precautions, at the outfit of my vessel, that no commodities should be put on board which could promote such a design, being determined, to adhere strictly to the articles, entered into by the courts of Great-Britain and Spain, respecting vessels, voyaging round Cape Horn. I accordingly shewed no colours, and as I kept my course, the fishermen, I presume, removed their signals.

On the eleventh day of June, at noon, I had got up the main, as high as the Isles Lobas le Mar*. I accordingly stood close in, within a mile or two of the shore, and then bore up for the isle, which we soon made, and got well in with it before it was quite dark, and then brought to, with our head to the Southward.

This isle, by my log, is sixteen leagues from the main, which, is a much greater distance, than is laid down, in most of the charts. My expectation was enlivened, in common with every one on board, by the opinion, that we should see some of our countrymen in the morning; and when we bore up at

break

---

* This isle was formerly the resort of the Buccaneers, but there is no fresh water on it

1793. of day, a confiderable quantity of tar was feen floating on the furface of the water; a circumftance, which ftrengthened our hope, that we fhould find a veffel refitting there.

I had fome intention of anchoring here myfelf, and having hove to, off the South Weft part of the ifle, I fent the chief mate to found for a dangerous rock under water, over which, the feas feldom or never breaks. It lays fomewhere, in the middle of the roads, and feveral whalers had ftruck on it; but I had not been able to procure the bearings of it. There was but little wind throughout the day, and the fhip fet confiderably to the Northward and Weftward, which opened the bay to us, when we were greatly difappointed, at not perceiving any fhip at anchor in it. However, before the boat returned in the evening, we faw a fail ftanding down on us, and it being hazy, as it generally is on this coaft, the boat had at one time miftaken her for the Rattler.

The chief mate returned on board by feven in the evening, and informed me, that he had not been able to difcover the rock, or to catch any thing but one turtle; but from the frefh carcaffes of feals which he had feen, he very reafonably fuppofed, that a veffel could not have left the ifland more than four or five days.

The

The fail, already mentioned, kept ſtanding towards us, and, as night advanced, ſhewed a light; at eight, being within a couple of miles of us, the whaling-maſter ſet out to board her, but, diſcovering on a near approach, that ſhe was a Spaniſh veſſel, he thought it right to return; I hauled on a wind for the night, as did the Spaniard, with a view of continuing together till morning; but the thick weather, which was not diſperſed on the return of day, prevented us from ſeeing each other again; nor did we perceive the land till ten A. M. when we found ourſelves ſet, during the night, within a few leagues of the Iſles of Lobas le Terra, which, in certain poſitions, bear ſuch a reſemblance to each other, that it was difficult to diſtinguiſh any difference between them: while, from the uncertainty of the currents on this coaſt, it might have been as naturally conjectured, that the current had ſet us as much one way as the other. As I had no inducement to beat back again, nor any probability of accompliſhing it, without taking a great offing, I continued on my courſe, but never failed to conſult with the whaling-maſter, before I ſhifted my ground.

The Iſle Lobas le Mar, is divided into two parts, by a ſmall channel, which will only admit the paſſage of boats, and where the tide is very rapid.

1793.   The Ifle Lobas le Terra, appears, towards the Eaftern point, to be much broken into fmall hillocks, while the land, or main near it, is low and vifible, only on a near approach.

During the fhort time I remained off thefe ifles, the weather was fo hazy, as to prevent my making any accurate obfervations concerning them.

June 16.   On the fixteenth of June, I reached Cape Blanco, the South Cape of the Gulf of Guiaquil, which is level land, of a moderate height, and, by feveral obfervations taken off it, I make it in Latitude 4° 8′ South, and Longitude 82° 20′ Weft. Off this cape, there is a ftrong, wefterly current, making out of the Gulf of Guiaquil; and afterwards, in croffing the gulf, I was in twenty-four hours, fet forty miles to the Weftward.

19.   On the nineteenth, I faw Point Saint Helena and Ifle Plata, where Admiral Sir Francis Drake divided his plunder. By feveral obfervations taken off the ifle, I place it in Latitude 1° 16′ South, and Longitude 82° 42′ Weft; and Point Saint Helena in Latitude 2° 0′ South, and Longitude 82° 20′ Weft.

The winds had now began to Weſtern on me, and knowing it, to be an object of the board of Admiralty, that I ſhould viſit the Gallipagoes Iſles, it became me to exert my beſt endeavours to do ſo, before I got further to the Northward; when, if the wind ſhould Weſtern more upon us, which it frequently does in this Latitude, I ſhould not have been able to fetch them.

1793.

On the ſame day I took my departure from Cape Saint Helena for Gallipagoes Iſles, for the reaſons already mentioned, the wind weſting on us; but, at thirty leagues diſtance from the coaſt, it returned to the South Eaſt quarter, and continued there, till we made the iſles. On the ſecond day, after we had left the coaſt, we fell in with a large flinched whale, which could not have been killed, more than three days. On the twenty-fourth, at four A. M., we made one of the Gallipagoe Iſles, bearing Weſt by North, ſix or ſeven leagues.

June 24.

In the courſe of our paſſage, we fell in frequently with ſtreams of current, at leaſt a mile in breadth, and of which there was no apparent termination. They frequently, changed the ſhip's courſe, againſt her helm, half the compaſs, although running, at the rate of three miles and an half an hour. I never

never experienced a similar current, but on the coast of Norway. The froth, and boil, of these streams, appear, at a very small distance, like heavy breakers; we sounded in several of them, and found no bottom with two hundred fathoms of line. I also tried the rate, and course of the stream, which was, South West by West, two miles and an half an hour. These streams are very partial, and we avoided them, whenever it was in our power. Birds, fish, turtles, seals, sun-fish and other marine animals kept constantly on the edge of them, and they were often seen, to contain large beds of cream-coloured blubber, of the same kind as those of a red hue, which are observable on the coast of Peru. The only seals, we saw, were in herds fishing, or in their passage, between the Gallipagoes, and the main. I do not affirm it as a fact, but as we saw no seals in my route back, and as the few, we killed there, were with young, I am disposed to conjecture, that the herds of them, just mentioned, were on their passage to whelp.

CHAP.

CHAPTER VI.

THE GALLIPAGOE ISLES.

AT day-break, 24th June, the land bore from West 10° South, to West 10° North by compass, having the appearance of two isles. It was my first design, to get round the Southernmost land, which was visible, and I accordingly hauled on a wind, but was induced, to alter my intentions, from a mistaken opinion, that I was further South than it afterwards appeared. I was led into this error, from having a North East current, during the two preceding days, setting at the rate of from twenty to thirty miles in the twenty-four hours. On rounding the North East point, which we passed at noon, the Latitude from observation was 40′ South, the East point bearing South East; and South West point South, 35° West. The soundings were ninety fathoms, and the distance, from the nearest land, eight or nine miles. The land,

1793.
June 24.

towards

towards the East, was covered with small trees or bushes without leaves, and very few spots of verdure were visible to us; a few seals were seen on the shore. The land rises at short intervening distances in small hills or hillocks, of very singular forms, which, when observed through a glass, and at no great distance from the shore, have the appearance of habitations, while the prickly pear-trees, and the torch thistles, look like their owners, standing around them. In other parts, the hills rise so sudden on the low land, that, having a small offing, they appear to be so many separate islands. About four miles off the North East end, there is a small islet, which is connected by a reef with the main isle: it is covered with seals, and the breakers reach some distance from the shore. The highest land, at this part of the isle, is of a very moderate height, descending gradually to the shore, which consists, alternately of rocks, and sand: some, of the rocky parts, being much insulated, they form winding inlets, of two or three miles in depth, and from one to two cables in breadth.

At the distance of two or three miles, to the Westward of the islet, I hove to, and sent the chief mate on shore to sound and land. At eight, P.M. he returned with green turtle and tortoises, turtle doves and guanas; but they saw no esculent vegetable,

vegetable, nor found any water that was sufficiently palatable to drink. He run four miles along the coast, at three quarters of a mile from the shore, without getting any soundings; at that length, found bottom at ten fathoms. This was near the distance we had fallen to leeward, from the time the boat had left us. I had sounded, several times, with the deep sea lead, at four or five miles from shore, and got no bottom, with one hundred and fifty fathoms of line. We stood off and on during the night, the wind being between the South and South East. At break of day, we discovered, that the current had taken a different direction, and had set us considerably to the Northward and Westward, and we could not fetch our situation of the preceding night. At noon, we were by observation, in latitude 37′ South.

I now thought it prudent to come to an anchor, in order to refresh the people, and to determine the situation of the isle. As we drew in with the shore, I kept the deep sea lead going, and at the distance, of about five or six miles, we obtained soundings, from thirty-eight, to thirty-six fathoms, which continued to diminish, till we were within a mile of the shore, when we got into nineteen fathoms water, fine sand bottom, and near the center of the isle; in which depth we came to anchor.

The land forms a kind of bay, the extremes of which bore, from South 43° West, an high bluff; to East 34° North, a low point; there is a distant high rock, off the South West point, West 33° South, which lays off the East part of a deep commodious bay. South by East of us, was a small bay, formed by two rocky points; in the East part of which, was one of those small creeks, already mentioned. I sounded round the ship with two boats, as well as between us, and the shore: here we found a good bottom, the soundings increasing or decreasing, as we distanced or neared the land.

Two boats now landed abreast of the ship, and the crews dividing, took the separate courses of East and West, in search of water and vegetables: a third boat I sent off to the large bay, which is distinguished by the high rock, on a similar pursuit, but they all returned in the evening, without having attained the objects of their search. The boat from the West, had found an uncommon kind of sand; we supposed it, from its weight, to contain some kind of ore, and which we afterwards found, to be small topazes.

This isle is of a moderate height, the highest parts being to the Westward. All the North side descends gradually to the sea, forming low points. Many parts are well wooded, but as it was winter, there was no appearance of verdure,

but

but from the evergreen trees and plants, such as the box and the prickly pear, with the torch thistle, and the mangrove. The middle of the isle is low land, and at a very small distance has the appearance of being divided into two parts, particularly on the South side. On the Western part of the bay, in which we anchored, the land is barren and rocky; in some parts, it has the appearance of being covered with cinders; and in others, with a kind of iron clinker, in flakes of several feet in circumference, and from one to three inches thick: in passing over them, they found like plates of iron: the earth is also frequently rent in cracks, that run irregularly from East to West, and are many fathoms deep: there were also large caves, and on the tops of every hill, which we ascended, was the mouth of a pit, whose depth must be immense, from the length of time, during which, a stone, that was thrown into it, was heard. Many of the cavities on the sides of the hills, as well as on the level ground, contained water, but of such a brackish taste, as to render it unfit to be drank. In most of them, there were considerable flocks of teals, which were by no means shy, and were easily caught: they are of the same kind as those known in England.

This island contains no great number, or variety, of land birds, and those I saw, were not remarkable for their novelty

novelty or beauty: they were the fly-catcher and creeper like thofe of New Zealand; a bird, refembling the fmall mocking bird, of the fame ifland; a black hawk, fomewhat larger than our fparrow hawks, and a bird of the fize and fhape of our black-bird. Ringdoves, of a dufky plumage, were feen in the greateft number : they feldom approached the fea till fun-fet, when they took their flight to the Weftward, and at fun-rife returned to the Eaftward; fo that if there is any water on the ifle, I fhould fuppofe it would be found in that part. Befides, it is the higheft land, and a fmall quantity of water, lodged in the hollow of a rock, would fupply thefe birds for a confiderable time. My fecond vifit, to thefe ifles confirmed, my fuppofition, as fmall oozings, were then found, at the foot of two or three hills, which may be occafioned by pools of rain water collected on the tops of them, as is frequently feen on the North Weft coaft of America. An officer and party, whom I fent to travel inland, faw many fpots, which had very lately contained frefh water, and about which, the land tortoifes appeared to be pining in great numbers. Several of them, were feen within land, as well as on the fea coaft, which, if they had been in flefh, would have weighed three hundred weight, but were now fcarcely one third of their full fize.

I was

I was very much perplexed, to form a satisfactory conjecture, how the small birds, which appeared to remain in one spot, supported themselves without water; but the party on their return informed me, that, having exhausted all their water, and reposing beneath a prickly pear-tree, almost choaked with thirst, they observed an old bird in the act of supplying three young ones with drink, by squeezing the berry of a tree into their mouths. It was about the size of a pea, and contained a watery juice, of an acid, but not unpleasant, taste. The bark of the tree, produces a considerable quantity of moisture, and, on being eaten, allays the thirst. In dry seasons, the land tortoise is seen to gnaw and suck it. The leaf of this tree, is like that of the bay tree, the fruit grows like cherries, whilst the juice of the bark dies the flesh a deep purple, and emits a grateful odor: a quality in common with the greater part of the trees and plants in this island; though it is soon lost, when the branches are separated from the trunks, or stems. The leaves of these trees also absorb the copious dews, which fall during the night, but in larger quantities at the full and change of the moon; the birds then pierce them with their bills, for the moisture they retain, and which, I believe, they also procure from the various plants and ever-greens. But when the dews fail in the summer season, thousands of these creatures perish; for, on our return hither, we found great numbers dead in
their

their nests, and some of them almost fledged. It may, however, be remarked, that this curious instinctive mode, of finding a substitute for water, is not peculiar, to the birds of this island; as nature has provided them with a similar resource in the fountain tree, that flourishes on the Isle Ferro, one of the Canaries; and several other trees and canes, which, Churchill tells us in his voyages, are to be found, on the mountains of the Phillipine Islands.

There is no tree, in this island, which measures more than twelve inches in circumference, except the prickly pear, some of which were three feet in the girth, and fifty feet in height. The torch thistle, which was the next in height, contains a liquid in its heart, which the birds drank, when it was cut down. They sometimes, even extracted it from the young trees, by piercing the trunks with their bills.

We searched with great diligence for the mineral mountain, mentioned by Dampier, but were not so fortunate as to discover it; unless it be that, from which the heavy sand or small topazes were collected, and of which, I ordered a barrel to be filled, and brought it away.

The great rock, bearing from our anchoring place, South 43° West, makes the East point of a large bay, in which, I anchored,

anchored, at our return. The winds that prevailed, while I 1793. lay here, were from South, South East, to South, South West, always moderate weather, but the tide runs very strong, particularly the flood, which comes from the Eastward: so that we were never wind rode; the ebb returns the same way, but not so strong; it is high water here, at the full and change of the moon, at half past three, and its rise twelve or thirteen feet. I place this isle between Latitude 45′ South, and 1° 5′ South, and Longitude 89° 24′, and it bears from Cape St. Helena, West 5° North, by compass, one hundred and thirty-five leagues. It lays in a North East and South West direction; and its greatest extent is thirteen leagues in length, and ten miles in breadth.

The various kinds of sea-birds, which I had seen on the Coast of Peru, we found here, but not in equal abundance. There were also flamingos, sea-pies, plovers, and sand-larks: The latter. were of the same kind, as those of New Zealand. No quadruped was seen on this island, and the greatest part of its inhabitants appeared to be of the reptile kind, as land tortoises, lizards, and spiders. We saw also dead snakes, which probably perished in the dry season. There were, besides, several species of insects, as ants, moths, and common flies, in great numbers; as well as grass-hoppers, and crickets.

1795.   On the shore were sea guanas and turtles; the latter, were of that kind, which bears a variegated shell. The guanas are small, and of a sooty black, which, if possible, heightens their native uglinefs* Indeed, so disgusting is their appearance, that no one on board could be prevailed on, to take them as food. I found the turtles, however, far superior to any I had before tasted. Their food, as well as that of the land tortoise, consists principally, of the bark and leaves of trees, particularly of the mangrove, which makes them very fat; though, in rainy seasons, when vegetation is more general, their food may be of a more promiscuous nature. The green turtles are extremely fat, and would produce a large quantity of oil. Their shell is also very beautiful; and if that should be an article of any value, a small vessel, might make a very profitable voyage, to this place. The land tortoise, was poor at this season, but made excellent broth. Their eggs are as large, and their shell as hard, as those of a goose, and form a perfect globe. Their nests, are thrown up in a circular form, and never contain more than three eggs, which are heated by the Sun,

an

* The sea guana is a non descript: it is less than the land guana and much uglier, they go to sea in herds, a fishing, and sun themselves, on the rocks, like seals, and may be called alligators, in miniature.

an hole, being so contrived, as to admit its rays through its daily course. The shell is perfectly smooth, and when highly polished, receives a beautiful and brilliant black.

We saw but few seals on the beach, either of the hairy or furry species. This circumstance, however, might be occasioned, by its not being the season for whelping; as those, which were killed by us, had some time to go with young; but a few hundreds of them, might at any time be collected without difficulty, and form, no inconsiderable addition, to the profits of a voyage.

Dampier mentions, that there is plenty of salt to be obtained here, at this season, but I could not find any; though that article does not appear to be absolutely necessary; as the skins will be more profitable, by drying and cleaning them, and then taking them to a China market; as I managed with the otter-skins, which I collected in a former voyage.

The rocks are covered with crabs, and there are also a few small wilks and winkles. A large quantity of dead shells, of various kinds, were washed upon the beach; all of which were familiar to me; among the rest, were the shells of large cray-fish, but we never caught any of them alive.

1793. On several parts of the shore, there was drift-wood, of a larger size, than any of the trees, that grow on the island: also bamboos and wild sugar canes, with a few small cocoa nuts at full growth, though not larger than a pigeon's egg. We observed also, some burnt wood, but that might have drifted from the continent, been thrown over-board from a ship, or fired by lightening on the spot.

The deep-water fish, were of every kind, that is usually found, in the tropical Latitudes, except spermaceti whale, and of them we saw none, but sharks were in great abundance.

The dip of the needle I found here to be at $84°$, and the variation of the compass $8° 10'$. The thermometer was never higher than $73\frac{1}{2}$, and in the morning, evening and night, it was below summer heat in England. I consider it as one of the most delightful climates under heaven, although situated, within a few miles of the Equator. The barometer generally stood at 29-8-4. The evening, night, and morning, were always clouded; and during the nights, there generally fell, as heavy dews, as off the main.

Every one was charmed with the place; but, as all apprehenſions of the ſcurvy or any other diſeaſe was at an end, and we had got a large proviſion of turtle on board, the anxiety of my people, to commence the fiſhery, in which, they all had a proportionate intereſt, began to ſhew itſelf; nor was I diſpoſed to check their ſpirits, or delay their wiſhes; being well aſſured, that they would be overjoyed to return hither, at no very diſtant period, when I ſhould have an opportunity to viſit the reſt of theſe iſlands.

1793.

On the twenty-eighth of June we weighed anchor, and ſailed round the Eaſt point, with a view of beating a ſmall diſtance to the Southward, in order to determine the particular iſle, we had viſited, according to the deſcription of the Buccaneers and the Spaniſh map, but my endeavours were not ſuccefsful. While we were at anchor, it was ſuppoſed, that we ſaw land in the North Weſt, at the diſtance of fourteen or fifteen leagues; but this was by no means aſcertained; though, according to Dampier, moſt of the iſles ought to have been in ſight of us, by allowing the difference of a few miles of Latitude between us and him.

June 28.

On the firſt of July, we ſaw a ſmall iſle which I beat up to; and, taking obſervations within a few miles of it, place

July 1.

it in Latitude 1° 24′ South, and Longitude 89° 47′ West. It bears, from the East point of the isle, before which we had anchored, South, distance five leagues, and lays in the direction of North, North West, and South, South East, and may be fourteen miles in extent. The side we saw, resembles the East point of the large isle, but is enlivened with an higher degree of verdure: we also saw a greater number of seals, off this, than off the other island. I do not hesitate to consider it, as the Southernmost and Easternmost of the Galapagoe Isles. In the accounts of Wood, Rogers and others, the Spaniards are said to be acquainted with an island in the Latitude of 1° 16′ South, which has plenty of water on it. This may be true during a rainy season, or for some time after it; but I am not in the habit of giving an implicit faith to Spanish accounts.

As I could not trace these isles, by any accounts or maps in my possession, I named one Chatham Isle, and the other Hood's Island, after the Lords Chatham and Hood.

CHAPTER VII.

PASSAGE FROM THE GALAPAGOE ISES, TO ISLE COCAS.

FROM the Southernmoſt Galapagoe Iſle, we ſtood over again for the main, keeping between the Latitude of 2° South, and the Equator, and had a ſtrong Eaſterly current conſtantly againſt us: but it was not ſo perceptible, as on our paſſage from the main, although we fell in with ſeveral beds of cream-coloured blubber: we did not, however, ſee ſo many ſmall fiſh, birds, or ſeals; of the latter, we only ſaw two, and they were not at any conſiderable diſtance from either iſle or main. Porpoiſes and black fiſh were continually around us, with a few albecores and bonettas.

1793.

The winds were much the ſame, as on my paſſage to the Galapagoes, blowing ſteadily from between the South and Eaſtward, after twenty-four hours ſail from the iſles;

and

1793. and, when within the fame diftance from the main land, they inclined to the Weftward: the weather was generally cloudy, and fometimes accompanied with an heavy, South Weft fwell, and at the change and full of the moon, with a drizzling rain.

July 10. On the tenth of July, P. M., we faw the Ifle of Plata, bearing Eaft North Eaft, nine or ten leagues, and, on the following day, in the morning, we faw fpermaceti whales, but did not get up with them until the afternoon, the Ifle Plata bearing Eaft by South, at the diftance of five or fix leagues. One of them was ftruck, but the boat was accidentally ftove to pieces, and we beat off for feveral days, but never got another view of them.

The Ifle Plata* had been a favourite place of refort to the Buccaneers, it being moft conveniently fituated to watch the Plata fleets to and from Lima; but all traders, either to or from the coaft of Mexico, or between Panama and the coaft of Peru, make the land a little to the Northward of it. If we may believe the Buccaneers, this ifland has plenty of water and turtle, and abounded with goats, till the Spaniards deftroyed them. The watering and anchoring places are faid to be on the Eaftern fide, in a fmall fandy bay, half a mile from the fhore, in eighteen or twenty fathoms water.

It

---

* So named by the Spaniards, from Admiral Sir Francis Drake dividing his plunder at it.

It is of a moderate height, and of a verdant shaggy appearance, from the large bushes or low trees that cover it. Its length is from six to seven miles; and the Western side is an entire cliff of an inacceffible appearance. A few small islets appear off the South end of it.

In a war with Spain this island would form an excellent station, as well as a place to look out and accommodate the sick, as it lies four leagues from the nearest main land, which is Cape Lorenzo. A ship getting in there, when it was dark, would not be discovered, if her sails were handed, the land being much higher than her mast head; unless the people on board betrayed her situation by some act of indiscretion, as making too much fire, the smoke of which might discover them. It is true that a vessel might escape by keeping an offing; but in so fine a climate as this, the long boats might form a chain to the Galapagoes, which is as far West as any ships are known to pass.

We continued a very affiduous search up the coast for whales, carrying an easy sail by day, and laying too at night, with an hourly expectation that we should fall in with them; but no whales shewed themselves, except some of the humpbacked species.

On

1793.
July 16.

On the sixteenth, at noon, off Cape Paſſado, the land being inviſible from the hazy weather, we were, by obſervation, in fifty-ſeven miles South. On the ſame day, we gave chace to, and came up with, a Spaniſh ſnow, from Acapuleo to Lima, from whom we procured ſome freſh beef, and two cocks; for which we returned a few bottles of wine and porter, with ſome ſweet-meats, the maſter being ſick.

18.

On the eighteenth, the weather became clouded and threatening, and I was every hour expecting to fall in with the heavy rains, which happen on the coaſt of Mexico, from November to July. The air alſo became hot and ſultry, and we had frequent ſhowers of rain. The thermometer now roſe to 80°, and we may be ſaid to have felt, at every pore, that we had left the moſt delightful climate in the world, to encounter the parching airs of the torrid zone *. The ſeals and birds, which are inhabitants of the frigid zone, but which I have ſeen as far South, as 70°, appearing to be delighted on the coaſt of Peru, as if inſtinct had forbidden them to venture no farther, now left us.

At this time it became neceſſary to determine, what route we ſhould take, whether we ſhould return to the Southward, or proceed to the Northward; but, as the

whaling

---

* On the coaſt of Peru it never rains.

whaling mafter and mates were in favour of a Northern Latitude, it was foon refolved to take our departure from Cape Paffado; which, from feveral obfervations, I place in Latitude ten miles South, and Longitude 82° Weft.

I now ftood acrofs the gulf, and, on the twentieth of July, fell in with the Ifle Malpelo; I had no defcription of this place, and I was not induced from its name, which fignifies bald head, to expect any advantage from it. I calculated its Latitude to be 4° 20′ North, and its Longitude 80° 45′ Weft, diftant from Cape Paffado eighty-fix leagues. It is a barren, and high, perpendicular rock, which may be feen, in clear weather, at the diftance of twenty leagues. A fmall quantity of green mofs, and a few dwarf bufhes, which grow in its cracks or gullies, afford the only verdure that it poffeffes: It is furrounded with iflets, and the whole may extend about nine or ten miles from North to South. The center, of this ifland, bears a refemblance, in feveral points of view, to the crown of an head; and its being barren, accounts naturally enough for the name, which the Spaniards, have beftowed upon it. Had I feen any feals off this place, I fhould have confidered it as a good fituation for them.

1793.

The Ifland of Malpelo, can be of no ufe, but as a place of rendezvous; it is furrounded, as it were, by a ftrong current, having much the appearance of breakers, which, fetting into the gulf and being accompanied by light winds, with thick and hazy weather, I did not think it deferving of any further attention. We tried the current and found it to fet North Eaft by Eaft, by compafs two miles and an half in the hour.

July 25.

From the Ifland Malpelo, we ftretched away, to the Weftward for Ifle Cocas, which we made on the twenty-fifth at midnight. The whole of the paffage thither, we had threatening, fqually and fhowery weather, with inceffant and heavy rain, and, at intervals, thunder and lightning: we had a fhort, irregular head fea, with winds from South, South Weft, to Weft South Weft. Porpoifes accompanied us in great numbers; and as we approached the Ifle Cocas, there appeared large flights of boobies, egg-birds, and man of war hawks. We alfo faw a fin-back whale, and two grampufes, with innumerable bonettas, dolphins, and albecores.

At break of day, the weather was thick and rainy, and, though the land was covered by the fog, we dif-
cerned

cerned several islands that lay around it. When we had got within four or five miles of the North East end, I sent a boat away with the chief mate, to search for an anchoring place; though, at times, I could not see the jib-boom end, so thick and frequent were the showers. At noon, the boat returned, having been in a bay near the North end of the isle, which was small, and open to the North East, with great depth of water, within three quarters of a mile of the shore. As this description did not answer to that of Mr. Wafer's bay, I stood in to examine it, as I could not have ventured to anchor in deep water, with a crippled windlass that occupied two hours, in a start calm, to heave in nineteen fathoms of cable: besides, the tide, which I found afterwards setting on both points of the bay, was so strong, that if the boats had not been very ready, the ship must have gone on shore; and, if in such a situation, there had been an anchor to heave up, it must have been cut away. I therefore ordered the boats to examine more to the Westward, and they accordingly discovered Mr. Wafer's harbour*. The land of this

---

Extract from Mr. Wafer's Voyage, who was at Anchor in this Bay, in 1685.

* Some or other of our men went on shore every day; and, one day among the rest, being minded to make themselves very merry, they went on shore, and cut down a great many cocoa trees, from which they gathered the fruit, and drew

about

this island is high, but that, on the West side, is the highest and presents itself in the form of a round hill. The Eastern side appears to be much broken, the land sloping in most parts, abruptly to the sea, but in others, presenting bold and perpendicular cliffs. There are also many surrounding islets whose tops are generally covered with trees; but the soil nevertheless is shallow, and the lower part is, as if it were a ring of white barren rock, down to the surface of the sea.

The main island does not appear to possess a spot, where trees can grow, that is not covered with them, or some kind of bushy plant, which when blended with the barreness of intervening rocks, produces a picturesque effect; while the streams, that pour down from their various fountains to the sea, greatly heighten the beauty of the scene. It is Otaheite on a small scale, but without the advantage of its climate, or the hospitality of its inhabitants.

Here

---

about twenty gallons of the milk: then they all sat down, and drank healths to the King, Queen, &c. They drank an excessive quantity, yet it did not end in drunkenness; but, however, that sort of liquor had so chilled and benumbed their nerves, that they could neither go nor stand: nor could they return on board the ship, without the help of those, who had not been partakers in the frolic; nor did they recover it under four or five days time.

Here are two anchoring places at this island; one, a small bay, near the North point of the isle; but the anchorage is in deep water, within three-quarters of a mile of the shore, from whence the bottom deepens almost immediately, to no soundings at sixty fathoms. It is also entirely open to the Northerly wind; but as Captain Vancouver anchored here after I left it, a more exact description may be expected from the promised publication of his voyage. I found the prevailing wind to be to the Southward and Westward; but, it often varied; and I had it frequently blowing strong from North East and North. The other bay, or harbour, is three miles to the Westward and Southward of the North point, and is easily known by a small rugged, barren rock, about the size of a large boat, bearing West of the body of the bay, about five or six miles: The bay also lies East and West; its greatest depth is not two miles, nor is it one in breadth: but I would not venture into it, in a vessel of more than two hundred tons. Its anchorage is from seven to fifty fathoms, and is nearly sheltered from all winds; this bay is also preferable to the one at the North point, because the shore of the first is steep; while that of the latter, consists of a beautiful valley and sandy

beach,

beach, where cocoa trees appear in greater numbers, than I have seen in any other place. There is also a rivulet of water eighteen or twenty feet in breadth, which is supplied from a bason, one mile distant within land, in which our crew, to avoid the sharks, went and bathed. Although this bay is so small, it is very convenient, and as secure, as the anchoring places generally are, which are not entirely sheltered. Its principal inconvenience arises from the constant rains; as out of the four days we were beating off it, it rained during three of them, in the offing, and sometimes with heavy storms of lightning and thunder. Those, who were on shore, experienced an equal continuance of the wet weather; and so thick was the rain, that, for eight hours together, we have not been able to see twice the length of the ship: but this may not be the case at all seasons. The woollen clothes of those who went on shore, which, had been particularly moist from perspiration, and were hung on the bushes to dry, were soon fly-blown, in the different parts that had stuck nearest to the body, and covered with maggots. Should a vessel touch here to recover her sick, or to water, or to wait any time, fire would remove the flies; and, as no tent would be sufficient to keep out the water, I would recommend the erection of an house, wood being in great plenty, and at hand, with cocoa tree leaves in abundance, to thatch it. I saw no plant,

bush

bush, or tree, but such as are quite familiar to my eye; they chiefly consisted of the mangrove, the cocoa nut, and cotton tree.

Fish were in great abundance, but would not take the bait, which we attributed to the great number of sharks off this island. Some of them followed the boat until the water left them almost dry: those we caught, were full of squid and cray-fish, as were the porpoises which we struck. These were innumerable, and we took them, whenever we pleased. Eels are plenty, and very large: we caught several of them among the rocks, as well as some toad fish. Shell-fish, were scarce, though we collected very large limpets, of a new kind, and a few dead conches. The latter were seen in great numbers on the beach, and mostly inhabited by the Diogenes crab. Common land crabs were in great plenty, and sea-birds of every kind, common to tropical Latitudes, in the Atlantic, were in great abundance here; particularly the Saint Helena pigeon, and white-headed noddy. They all perched on trees, like land-birds; and, at a small distance, gave the tree on which they sat, the appearance of being covered with white blossoms. Of the land-birds, we saw

some,

some, which resembled the thrush and blackbird, in shape, colour, and size, with a few herons and a variety of smaller birds.

The tide must be an object of particular attention, in anchoring at, or sailing from, this place: it ebbs and flows from sixteen to eighteen feet, perpendicular, and, from the observations made by myself and the officers in the boats, it flows seven, and ebbs five hours; the ebb setting to the Eastward, and the flood to the Westward: but the flood runs not near so strong, as the ebb, which runs at the rate of four or five knots an hour. The time of weighing and anchoring must also be attended to, as both sets are right on the points of the bay; and, if its rise and falls are regular, it will be high water at full and change, at four, A. M.

The rats, which are numerous, in this island, exactly resemble the common rat in England, and were, probably, left here by the Buccaneers. As we found their nests in the top of most of the trees which we cut down. I am disposed to conjecture, that this is a very humid spot, at all times and seasons.

I was

I was much disappointed, at not being able to procure turtles; for we saw but two, and they escaped us. That there should be so few turtles here, must be owing to the great number of sharks that infest the coast, or the chilling rains, which destroy the eggs, when laid on the shore, which, in itself, is very favourable to their becoming productive. There is as fine and soft a beach, as I ever saw, and there are few vessels, but might lay a-ground on it, and repair and clean their bottoms. Whoever may, hereafter, wish to anchor in this bay, will do well, to come round the South and West points of the isle, and hug the South point of the bay, close on board, and when in the bay, to moor head and stern.

We were much wearied, during the four days, we passed off this island, and prepared to quit it. We therefore took on board, two thousand cocoa nuts; and, in return, left on shore, in the North bay, a boar, and sow, with a male and female goat. In the other bay, we sowed garden seeds, of every kind, for the benefit and comfort of those who might come after us. I also left a bottle tied to a tree, containing a letter. Over it, I ordered a board, with a suitable inscription, which Captain Vancouver thought proper to remove, when he anchored at this isle, some time after me. The letter gave only an

1793. account of my arrival and departure. Having made the neceſſary arrangements, we ſet ſail for the Northward.

Iſle Cocas lays in a North Eaſt and South Weſt direction; its greateſt length does not exceed twelve miles, nor breadth four miles.

It may be proper to remark, in this place, that, in all parts of the Eaſt Indies, a vinegar is made of the milk of the cocoa nut, equal to our ſtrong white wine vinegar. I am unacquainted with the particular procefs, but am difpofed to think it at once ſhort and ſimple. The old cocoa nut left in water for two hours, and then ſtrained, produces a liquid, in colour and taſte, little inferior, if not equal, to ſkim milk, which removed all ſcorbutic complaints from among the crew, and preſerved them in health, for many months.

CHAP.

CHAPTER VIII.

ROUTE FROM ISLE COCAS, TO THE COAST OF MEXICO; AND FROM THENCE, TO THE ISLES SOCORO, SANTO BERTO AND ROCKA PARTIDO.

THE Isle Cocas, was the fartheft point to the Northward, to which it was recommended to me, by the Board of Admiralty, to extend my refearches; but an anxiety and emulation to afcertain every part, and defcribe the whole furface of the feas, where the fifhery could be extended, would have enlarged the circle of my voyage, if my ftock of provifions and ftores had been fufficient for fuch a defign: I was therefore obliged to check my intentions, having, for the reafons above-mentioned, time only to examine as far as 24° 0′ North, on the coaft and gulf of California, down the coaft of Mexico to Ifles Socoro, Santo Berto and Rocka Partido, and off the North Weft point of the gulf of Panama.

1793.   This was an undertaking that few, who had suffered as I had done, from the yellow fever, in the prisons of New Spain, as well as from all the horrors of a rainy season on that coast, would have encountered; but I was persuaded, within myself, that there must be plenty of spermaceti whales on this coast; and it was very evident, that if successful in killing them in the rainy season, it must be much more easily done in the dry season. At all events, I was determined to make the experiment.

On leaving the Isle Cocas, we stood away to the Westward and Northward, in the hope of, avoiding the rain in some degree, by keeping at a small distance from the land.

August 1.   On the first of August we were in Latitude by observation 9° 2′, and Longitude corrected 90° 0′ West. We bettered our weather greatly; but the heat was almost intolerable; the thermometer standing at 86°, and the barometer at 29-7-8; the wind now began to vary to the Eastward.

3.   On the third of August our Latitude was 9° 30′ and Longitude corrected 89° 44′. The bad weather returned and continued with frequent tornados and heavy rain.

On

On the feventh of Auguft, we faw the famous burning mountain of Guata-mala. From that time, to our croffing the gulf of Tecoantepeak, and reaching point Angels in Latitude 16° and Longitude 100°, there was, for nine days little or no ceffation of calms, and the change that followed was a feries of tornados, torrents of rain and tremendous thunder and lightning, more violent than any I ever heard or faw on the coaft of Guinea, or off the capes of Virginia in North America. If there was any difference, in the fervid feverity of the feafon, during the twenty-four hours, it was in favor of the day; for in the night the lightning and thunder were moft violent. From fun-fet till fun-rife, the heavens were one entire flame, which was heightened, by the frequent explofions of the burning mountains. This awful and alarming ftate. of the weather, was accompanied with an almoft infufferable heat, and a fuffocating, fulphureous air. From the light airs, calms and variable currents, we had little hopes to fhift our fituation. Thus furrounded, as we were, with thefe oppreffive circumftances, and divided only by a few leagues acrofs the main from the bay of Honduras, it was impoffible to fupprefs an occafional wifh that we were there. A traveller that had vifited Peru or its coafts, (where the dews of bounteous Heaven fupply every call of rain, and where thunder and lightning are feldom or

never

1793. never known, and nature rests in perfect tranquility), would when here, naturally remark, that Providence had blessed the coasts of Peru, by exempting that country from all convulsions to be dreaded from the aerial elements, and doubly bestowed them as a curse on this; unless they are to be considered as a blessing, to impress the untutored Indian inhabitanr, " by feeing God in clouds and hearing him in the winds", with a due idea of his Almighty power. It is also to be hoped, that a native and resident in Peru, feels, sometimes, ideas of gratitude and thankfulnes towards his maker, for his goodness. We founded frequently, in the gulf, at twelve or thirteen leagues distance from the shore, and found no bottom with one hundred and fifty fathoms of line : but when in Latitude 14° 57', and at ten or eleven leagues distant, we got bottom, at one hundred and five fathoms, which was muddy.

August 19. On the nineteenth of August, when two papps over point Angels, bore North East, and our Latitude was 16° 13' North, we saw a large body of spermaceti whales, and though the spirits of my people, were in some degree depressed, by reflecting on the immense body of water over which we had sailed, the little success which we had hitherto experienced, and our being at least a seven months voyage from England, they were now elated, with all the eagerness of sanguine expectation. The boats accordingly gave chace, and soon came up

with

with the whales, though they were running faſt to the Southward, and appeared to be larger than any that had been ſeen by thoſe in purſuit of them. There being light airs, and calms alternately, the ſhip followed but ſlowly: the fiſhers ſtruck ſeveral whales, but were not ſo fortunate as to kill any of them.

The people in the boats, had now been away ſeven hours, and were ſo far diſtant, that the ſhips top-ſails, to them, were in the horizon; the day alſo was far advanced, and purſuing the whales, in the direction they were running, would be ſtill increaſing their diſtance, without a flattering hope, of ſaving the fiſh, if they killed them; ſeveral water-ſpouts were viſible in the horizon, accompanied by diſtant thunder and lightning, with a threatning ſky; all theſe circumſtances combined, obliged them, for ſelf-preſervation, reluctantly to give up the chace, and by the time they reached the ſhip, from the vaſt quantity of water they had drank, and the exceſſive heat of the weather, (which was in no ſmall degree increaſed by the fatigue undergone, and diſappointment occaſioned by their fruitleſs exertions) the whole of my crew were ſeized with a ſevere ſickneſs, and one of them was ſo cramped, that he would certainly have expired, if he had not almoſt inſtantly, on his return, been immerged in warm water. Another broke out in a violent raſh from head to foot, which, by his plunging

in

1793. in that ſtate into the ſea, was thrown into his head, and deprived him of his ſight for ſeveral days; I was very apprehenſive, that he would never recover it, but by placing him in warm water, frequently, in the courſe of the day, the raſh returned to his thighs and legs, and by degrees, his ſight was reſtored.

The hope of more favorable weather, and of better ſucceſs, in our commercial objects, induced me to remain cruiſing here ſixteen days; during which period, we ſaw whales, three different times, and killed three of them. One was a ſmall one, meaſuring 15 feet, which we hoiſted on board, and of which I made a drawing; its heart was cooked in a ſea-pye, and afforded an excellent meal. Theſe whales were very poor, having ſcarce blubber enough, to float them on the ſurface of the water, and when flinched, their carcaſes ſunk like a ſtone. They yielded altogether but fifteen barrels of oil.

The weather remained unpleaſant, there being ſcarcely any interval for the better, for twenty hours, with a ſtrong Southerly current of half a mile an hour. The whole crew had been, more or leſs, affected by the yellow fever, from which horrid diſorder, I was, however, ſo fortunate, as to recover them, by

adopting

adopting the method that I saw practised by the natives of Spanish America, when I was a prisoner among them. On the first symptoms appearing the fore-part of the head was immediately shaved, and the temples, and pole, washed with vinegar and water. The whole body, was then immersed in warm water, to give a free course to perspiration; some opening medicine was afterward administered, and every four hours, a dose of ten grains of James's powders. If the patient was thirsty, the drink was weak white wine and water, and a slice of bread to satisfy an inclination to eat. An increasing appetite was gratified by a small quantity of soup, made from the mucilagenous parts of the turtle, with a little vinegar in it. I also gave the sick, sweetmeats and other articles from my private stock, whenever they expressed a distant wish for any, which I could supply them with. By this mode of treatment, the whole crew improved in their health, except the carpenter, who, though a very stout, robust man, was, at one time, in such a state of delirium, and so much reduced, that I gave him over; but he at length recovered.

As the yellow fever seldom attacks any one twice, while he remains near the same place, my apprehensions were now confined to the scurvy and other incidental disorders;

but they were sufficient to quicken my anxiety, to find a place for refreshment, whenever it might be wanted. For though my crew were at present in good health and spirits, I had learned by my former expeditions, that there is no circumstance which operates more favourably on the temper and disposition of sailors in long voyages, (whenever they are attacked with those diseases to which they are so subject and of course most frequently dread) than the certainty of a port or harbour to which they may be taken; experience having also taught them, that the smell of the shore and change of sea diet, in general, remove the greatest part of their complaints*.

We brought plenty of cocoa nuts from Isles Cocos, and there was never wanting a fresh meal of turtle; for they were

---

* I do not pretend to any other medical knowledge, but such as I may have acquired, by some little reading on medical subjects, and the attention I was obligated to pay to the diseases and complaints of seamen, in the various voyages I have made, as it frequently became a nice point to judge, whether a man neglected his duty from idleness or sickness. I also paid particular attention to the practice of the different Indian nations, when an opportunity was afforded me, and from the circumstance of having no surgeon on board, it became a duty in me, to make part of my study, such an important subject, as the health of my crew; and I was so fortunate as to succeed in the applications I used, as to restore health through means, which the suggestions of the moment only dictated to me.

were in such numbers floating on the surface of the water, as to be taken whenever they were required. To this food, we may be said to owe the preservation of our healths, and the crew, in general, grew fat upon it.

Other voyagers have alledged, that living on turtle, causes the flux, scurvy and fever; I can first account for such a consequence, by its not being sufficiently boiled, or cooked in unclean utensils; and, secondly, every man who has experienced a long voyage, is well informed, that a sudden change of food, and particularly from an ordinary sea or salt diet, to an entirely fresh one, will produce the flux, sickness of stomach and other complaints. My method, to prevent such effects, was to allow the crew as much vinegar as they could use, and superintend myself the preparation of the seamen's meal. I used to taste the broth, in order to know if it was properly done, that it contained a sufficient quantity of pearl barley, and was duly seasoned by boiling with it salt beef or pork. I also ordered that the proportion of the salt meats cooked with the turtle, should be previously towed and freshened, and when the crew were tired of soup, I gave them flour to make their turtle-meat into pies, and, at other times, fat pork to chop up with it, and make sausages. But in most of their messes, I took

1793.
September. took care that so powerful an antiseptic, as sour crout, should not be forgotten.

For the reasons already mentioned, I determined to stretch off to the Westward, to search for Isle Socoro, Santo Berto, and Rocka Partida, but, although I thought it right to leave the coast for the present, I did not give up my opinion, that a whaling voyage might be made in the dry season, which would probably commence within less than two months; at the expiration of which time, I was determined to return. Nothing, indeed, would have deterred me from it, at present, if we had possessed sufficient wind to shift our situation, and keep the run of the fish, or clear weather, to ascertain the true Latitude and distance, from the land, at which, we fell in with them.

There were many ports near this, into which I wished to have entered, particularly the famous one of Guatalco, where Sir Francis Drake, got a bushel of money, out of one house, in 1579; and, in 1587, Sir Thomas Cavendish, possessed himself of great riches: but being naturally led to believe, that the above circumstances would not be forgot by the natives, and my ill treatment at Nootka, and St. Blas, being also fresh in my memory, I though it most prudent to give up, for the present, all ideas of going into any harbour on the Spanish coast.

On

On the ninth of September, in Latitude 17° 16′, and Longitude 102° 32′, we met with as irregular a fwell as I ever faw, off Cape Horn, accompanied with very changeable weather, faint lightning round the compafs frequent fhowers of rain, and light variable winds, blowing North Weft by Weft, round the compafs, to Eaft South Eaft, and continually fhifting till the 17th of September, at midnight; when, in a heavy fquall of wind from the North Weft by Weft, there fell as great a torrent of rain, as I had feen, with tremendous thunder and lightning, which I concluded was the forerunner of the equinoctial gale: on the 17th at noon, our Latitude was 18° 27′ North, Longitude, 109° 0′ Weft; thermometer 30°, barometer 29 6 4 ; at this time blowing a ftrong breeze, and unfettled weather, which, by the eighteenth, at noon, had increafed to a perfect ftorm, from the Weft North Weft, with a very heavy fea, that we could fhew little or no fail, till eight o'clock the fame evening; when the weather moderated, thunder, lightning, and rain ceafed, and the wind fettled in the Weftern quarter.

At day-break, on the twentieth, we faw the Ifland of Socoro: a number of thofe birds that generally follow the

1793. the spermaceti whales, as well as others, of a different species, accompanied us. At five o'clock in the evening, when we were within seven or eight miles of the shore, it being a moon-light night, I sent the chief mate to fish, sound for an anchoring place, and, if possible, to land, in order to discover what this island produced. We stood on and off during the whole night, and, at break of day, found that the current had set us considerably to the Southward and Westward. In the morning, we passed great quantities of pumice stone, and the sea was covered with small shrimps, the common food of the black whale. It being calm, or light winds all night, and the first part of the day, we did not get in with the shore, till two o'clock in the afternoon. We sounded within five miles of it, but found no bottom, with one hundred and fifty fathoms of line.

In the evening, the boat returned, when the mate informed me, that he had sounded off the lee-side of the isle, and could not find a place of safety for the ship to lay in, or a landing for the boat, except in a small cove, near the South point. They had caught a sufficient quantity of fish for all hands, consisting of a kind of cod, snapper, and silver-fish; and they might have taken more, but the sharks, which were very numerous, ran away with the hooks. On the island they had

had gathered a large quantity of beans, known, I believe, by the name of the Spanish broom: they also brought with them a confiderable number of prickly pears. As foon as it was light, I fent the boat, with cocoa nuts and garden feeds of every kind, which I caufed to be fown in the fmall cove, at the South point, and ftood with the ship off and on till they returned. In the afternoon, being within three or four miles of the cove, we got bottom, at forty-eight fathoms, fine fand. I then fent a boat, to found between us and the land, as well as to the Weftward, when bottom was found at ten fathoms depth, at half a mile from the fhore, to fifty fathoms, at three or four miles diftance.

By ten the next day, I had coafted the South and Weft parts of the ifle, and founded frequently, particularly in a fmall bay, at the North Weft, where we found good bottom, but it was expofed to the North Weft winds, which are reprefented to be the prevailing ones: though I found the winds, in general, Eafterly. The unfettled weather we had lately experienced, was fufficient to prevent my anchoring at this feafon, although in with the ifle, unlefs in a more fecurely fheltered bay, then I had as yet difcovered.

We

1793.   We saw Isle Santo Berto from the West end of this isle, bearing North 20° East. Having made Socoro and Santo Berto, by the Spanish manuscript chart, which I procured, while a prisoner at St. Blas, and got a sufficient store of beans and prickly pears; I streched away to search for Rocka Partida and St. Thomas's, by the same chart. Two of the crew were affected with a violent purging and vomiting, from eating too much of the fruits just mentioned. It lasted twenty-four hours, and, in the end, proved beneficial to them. Indeed, we were all in perfect health, except the second mate, who had a lameness and contraction in one of his knees, and had been in an ailing state, ever since we left Rio Janeiro.

Sept. 24.   On the twenty-fourth, at nine, A. M. we saw Rocka Partida, on our weather bow, which had the appearance of a sail. By four o'clock, we worked up with it, and found it a dangerous barren rock, laying North, North West, and South, South East, by compass. Its greatest length, is fifty or sixty fathoms: and its breadth, about twenty-five or thirty: both ends are fifteen or twenty fathoms in height. The North West end is forked; the South and East end, is like a ragged hay-cock. The two heights are separated by a ragged saddle, that rises about three or four

four fathoms from the surface of the sea, and is nearly perpendicular. On sounding all around, at a boats length, we had thirty-five fathoms; and, at half a mile distance, fifty fathoms; and then no bottom, with an hundred fathoms of line. It shews itself, on every bearing of the compass, from a small to a great distance, like a sail under a jury-mast. This rock is situated in Latitude 19° 4′ 30″, and Longitude, by observation of Sun and Moon, and chronometer, corrected, 111° 6′ 30″, bearing from the South West end of Isle Socoro, West 15° North, by compass; distant forty-eight miles: the variation, 7° East. I leave the further descriptions of Isles Socoro and Santo Berto, to my return and anchoring at the first mentioned isle, when I had a better opportunity, and more time to make remarks.

At Rocka Partida was a prodigious quantity of fish, but we caught only few, as the sharks destroyed our hooks and lines, and no one on board, but myself, had ever before seen them so ravenous. One of our men reaching over the gun-whale of the boat, a shark of eighteen or twenty feet in length, rose out of the water to seize his hand, a circumstance not uncommon at the Sandwich Isles, where I have seen a large shark take hold of an outrigger of a canoe, and endeavour to overset it. This was in some degree the case with our boat; a number of them continually seizing the steering oar, it became of no use, so that we were obliged to lay it in. The inhabitants of the rock were, as many man-of-war hawks as could find a resting place, and a few seals.

Having found the Isles Socoro, Santo Berto, and Rocka Partida, by my manuscript chart, I had every reasonable expectation of seeing also the Isle St. Thomas, which was discovered by a Spaniard, Diego Hurtado, in the year 1533, and by him placed in Latitude 21° 30'; and it was visited afterwards by Gaeten Bestrad, in the year 1542, who places it fifteen miles more to the Northward, than Hurtado; and by all the information I had collected, it lay a small distance to the Westward and Northward of Socoro.

I shaped my course for the situation in which it was placed in my chart; but when I had run the distance, I did not perceive any thing like land, nor any signs of my being near it, except the birds and seals which we frequently saw. I did not, however, entertain the least doubt of its existence, but concluded that I had missed it by sailing two much in a right line from Rocka Partida. The weather being too unfavourable for me to return to the coast of Mexico, I discontinued my search, for the present, after the Isle St. Thomas; and, from the quantity of whales frequenting the coast of California, as mentioned by Mr. Dalrymple, in his history of that country, as well as from the number seen by myself in my preceding voyage,

and

and the information I received from the Spaniards, while I 1793. I refided among them, I was determined to make a trial of fifhing there, till the fair weather came on to the Southward; which might reafonably be expected to begin at Cape Corientes, the latter end of October, or beginning of November. In my route to the coaft, I endeavoured to make Clipperton's Ifle from the beft accounts I poffeffed; but they differed fo widely in Latitude, that I was at a lofs where to look for it; and, as it was not in my defign to come this way when I failed from England, I had left behind me my manufcript chart of the feas, &c. laying North of Ifle Socoro, with all the information I had received from the Spaniards concerning them.

From the twenty-ninth to the thirtieth, we beat to Sep. 29-30. the Northward, in fhort tacks, with the hope of defcrying Clipperton's Ifle; we faw frequently man-of-war hawks, and at times a few folitary feals. As we had fome expectation of feeing land, every cloud that rofe in the horizon was declared, by the feamen, to be the object in fearch: but as I could not be perfuaded it was, I did not think proper to purfue the various momentary opinions which frequently were ftarted,

On the fourth of October, in Latitude 23° 15′, we October 4. made the coaft of California. The winds from the time of our leaving Socoro, blew from North North Eaft, to North North Weft, wefting as we made the land of California, with very pleafant weather, but fometimes cloudy. On our paffage we faw a few turtles, with killers, porpoifes,

1793. poises, and black-fish: the latter were innumerable as we approached the land.

October 12. We cruized off this coast till the twelfth, seeing only the kind of fish already mentioned, with the addition of some fin-back and hump-back whales; so that we had no inducement to remain there, after we had ascertained that the species of whale on this coast is of no value. Our cruizing ground was between the Latitudes 23° and 25°, and Longitude 112° and 113°, off a remarkable mountain near Cape St. Lazarus; to which I have given the same name: I make it to be in Latitude 25° 15′, and Longitude 112° 20′. To the South of it, is very low land, till within a few leagues of Cape St. Lucas, which makes the South point of California, when the land rises to such an eminence, as to be seen at the distance of twenty leagues: but the Cape itself is of a very moderate height. Though the weather was fair and pleasant, it was so hazy while we were on this low and dangerous coast, as to require a continual employment of the lead. We frequently got soundings with seventy fathoms of line at the distance of nine leagues from the shore.

I am ready to confess, that I was deceived respecting the species of whale which I saw when I was on on the coaſt before; and at this time the hump-back whale was ſo much larger than generally believed, and ſpouted in a manner ſo different from their uſual mode of throwing up the water, that the moſt experienced fiſhermen I had on board believed them to be black whale, and purſued them as ſuch; and I very much doubt whether that ſpecies of whale, which the Spaniards call the ſmall whale, is any other than black fiſh. This opinion was confirmed by a whaler, with whom I fell in company ſome time after. He had come down the coaſt of California, and boaſted of the number of ſpermaceti whales which he had ſeen. I was very much aſtoniſhed that, provided as he was for the purpoſe, he had not even attempted to kill one of them. But he ſoon ſatisfied my doubts on the ſubject: for being with me on board the Rattler, and ſeeing a ſhoal of black fiſh, he inſiſted that they were ſpermaceti whales. While I thus diſcovered his ignorance, I had reaſon to be ſatisfied with myſelf, in having been able to aſcertain, from the deck of my ſhip, the difference between theſe two ſpecies of whale, but this I muſt acknowledge, that black fiſh, in their feeding and mode of ſpouting, reſemble the ſpermaceti whale nearer than any other fiſh hitherto known.

On

1793.

On the twelfth at noon, Cape St. Lucas, the North Cape to the gulf of California, bore North twelve or thirteen leagues. I make this cape by the mean of a number of obfervations, of Sun, Moon, and Stars, in Latitude 22° 45′, and Longitude 110°. The fea, at this time, was almoft covered with turtles, and other tropical fifh. At four, A. M. we faw a large fpermaceti whale, which we ftruck and got faft: but night coming on, the irons drew, and it was loft. We cruized between the Cape Corientes, the South cape of the gulf of California, and the northernmoft of Maria Ifles, till the feventh of November, and faw great numbers of fpermaceti whales, fome of them the largeft we had ever feen, but we may be truly faid to be unfortunate, as we only killed two of them

Nov. 1.

Two of the crew, who complained of fome fcorbutic fymptoms, on the firft of November, were now growing worfe; and, as feveral others were apprehenfive of being attacked by this terrible diforder, it became neceffary for me to repair to fome port, where a proper atten tion might be paid to the invalid part of my people.

Our cruizing was generally at the diftance of from three to feven leagues to the Weftward of the Ifles Tres Marias, the largeft of which has been faid to have

have a good road, and to afford various articles of refreshment: but the French navigator, Monsieur Sauvage le Muet, who visited these isles in this month, in the year 1741, mentions, that his crew grew worse while he remained there.

The healthy season, which was now only beginning at St. Blas, situated in the mouth of the river St. Jago, at little more than twenty leagues from them, might not extend to those isles so early as November; and, in the bad season, at that place, it is not uncommon for six or seven of the natives to die in the course of a day, out of the small number of five or six hundred inhabitants. Besides, I could not help recurring, with many a melancholy thought, to the fate of my crew, in my former voyage, when we were captured by the Spaniards at Nootka, carried to St. Blas, and treated with the greatest inhumanity. I was determined, therefore, not to risk a second capture and imprisonment by the Spaniards, which would not have been improbable, if we had anchored at the Tres Marias: the launches from the royal dock at St. Blas, frequently visiting these isles, in order to get flax and lignum vitæ: nor have I the least doubt of their attempting it, if they

they had found me there in so capital a ship as the Rattler, and in so defenceless a state as she then was, armed with only two three-pounders, and half a dozen old musquets *.

The

---

* As there have been various misrepresentations of the real state and progress of the transactions between Don Martinez, commander of certain ships in the service of his most Catholic Majesty at Nootka Sound, and several trading vessels belonging to subjects of Great Britain, which threatened to produce a rupture between the two courts; and, as those misrepresentations may be hereafter repeated, I shall beg leave to give a fair and correct statement of those transactions, so far as I was unfortunately, involved in them: the rest of that unpleasant business is detailed at large, and accompanied by authentic documents, in the Appendix to the voyage of Captain Mears, published in London, 1790.

It is unnecessary upon this occasion, to have recourse to any occurrences in that unfortunate voyage, prior to the time when I appeared off Nootka, viz. the third day of July, 1789. At nine in the evening, when it was almost dark, we hailed a boat; and the persons in it desiring to come on board, their request was immediately granted. It proved to be a Spanish launch, with Don Estevan Martinez, commodore of some Spanish ships of war, then lying in Friendly Cove: we were visited at the same time by another Spanish launch, and the boat of an American ship. I had no sooner received Don Martinez in my cabin, than he presented me a letter from Mr. Hudson, commander of the Princess Royal Sloop, which was under my orders. The commodore then informed me, that the vessels under his command were in great distress, from the want of provisions and other necessaries; and requested me, in a very urgent manner, to go into port, in order to afford him the necessary supplies. I hesitated, however, to comply with this demand, as I entertained very reasonable doubts, of the propriety of putting myself under the command of

two

The Tres Marias, or the iflands, fo named by the Spaniards, off which we had been cruizing, are four in number, if the Ifle Saint John is included, which is not

more

---

two Spanifh men of war. The Spaniard obferving my unwillingnefs to comply with his requeft, affured me, on his word and honor, in the name of the King of Spain, whofe fervant he was, and of the Viceroy of Mexico, whofe nephew he declared himfelf to be, that, if I would go into port and relieve his wants, I fhould be at liberty to fail whenever I pleafed. He alfo added, that his bufinefs at Nootka was for no other purpofe, than merely to prevent the Ruffians from fettling on that part of the coaft, and that it formed a leading principle of his inftructions, as it was his private inclination, to pay all becoming refpect and attention to every other nation. I am ready to acknowledge that the ftory of his diftreffes, and the letter of Mr. Hudfon, which appeared to be deferving of credit, had very confiderable weight with me: befides, I was an officer in his Britannic Majefty's fervice; and might be, in fome degree, influenced by a profeffional fympathy. I therefore fuffered myfelf to be perfuaded to enter the harbour; and, as it was a calm, to let the Spanifh boats affift in towing the Argonaut into Friendly Cove; where we arrived by twelve at night and found an American fhip called the Columbia, riding at anchor, commanded by Mr. Kendric, and a floop of the fame nation, called the Wafhington, commanded by Mr. Gray; with two Spanifh fhips of war, called the Princeffa, and Don Carlos. The next morning, after I had ordered fome provifions and ftores for the relief of Don Martinez to be got ready, I went to breakfaft with him, in confequence of his invitation. After breakfaft he accompanied me on board my fhip, the Argonaut; I gave him a lift of the articles I intended to fend him, with which he appeared highly pleafed. I then informed him it was my intention to go to fea in the courfe of the day: he replied, he would fend his launch to affift me out of the harbour,

and

more than six miles distant from the Northernmost. There are also many small rocks, whose heads just rise above the water. All these islands are covered with wood, and

---

lay and that I might, on the return of the boat, send him the promised supply. The launch not coming so early as I wished, I sent one of the mates for her, but instead of bringing me the boat, I received an order from Don Martinez, to come on board his ship and bring with me my papers. This order appeared strange, but I complied with it, and went on board the Princessa. On my coming into his cabin, he said he wished to see my papers: on my presenting them to him, he just glanced his eyes over them, and although he did not understand a word of the language in which they were written, declared they were forged, and threw them disdainfully on the table, saying at the same time, I should not sail until he pleased. On my making some remonstrances at his breach of faith, and his forgetfulness of that word and honour which he had pledged to me, he arose in an apparent anger, and went out.

I now saw, but too late, the duplicity of this Spaniard, and was conversing with the interpreter on the subject, when having my back towards the cabin door, I by chance cast my eyes on a looking-glass, and saw an armed party rushing in behind me. I instantly put my hand to my hanger, but before I had time to place myself in a posture of defence, a violent blow brought me to the ground. I was then ordered into the stocks, and closely confined; after which, they seized my ship and cargo, imprisoned my officers, and put my men in irons. They sent their boats likewise to sea and seized the sloop Princess Royal, and brought her into port, for trading on the coast.

It may not be amiss to observe, that the Spaniards consider it contrary to Treaty, and are extremely jealous, if any European power trades in those seas, but this cannot justify Don Martinez, who, not content with securing me and my people, carried me from ship to ship, like a criminal,

rove

lay between the Latitude of 21° 15′ and 22° and Longitude 107° West. The center isle is the largest; the Northern-most, which is named Saint John, is low and tabling, but

---

rove a halter to the yard-arm, and frequently threatened me with instant death, by hanging me as a pirate. This treatment, at length, nearly cost me my life; and threw me into so violent a fever, that I was delirious for several days: After recovering, I was sent in my own ship prisoner to St. Blas a Spanish port in the Gulf of California. On my passage thither, I was confined in the Mate's-Cabin, (a place not six feet square) for two and thirty days, with a scanty supply of miserable provisions, and a short allowance of water. The British part of my ship's company, with two officers, were confined in the sail room with their feet in irons, and kept in a state too shocking to relate, and which decency forbids me to describe. In going into the harbour, the Spaniards ran the ship aground and damaged her bottom. On landing, few of my people had any change of clothes, for the Spaniards had broke open their chests and plundered them; however, when under the care of the Governor of St. Blas, we were better treated, being permitted to walk about the town, in charge of a guard of soldiers, and allowed sufficient provisions. About this time the Princess Royal and crew arrived, and shared the same fate. Soon after, under a promise that our detention could not be long, they persuaded us to heave down and repair the Argonaut, new copper her bottom, and fit new rigging. The idea of release stimulated us to work on the ship with great alacrity, so much so, that our exertions threw several into fevers; and on the vessel being nearly ready, the Governor threw off the mask, informing us she was to be employed for their use, and laughed again at our credulity. This treatment, added to little thefts committed on us with impunity, worked on the minds of the sickly part of the crew, several of whom took it to heart and died, and one destroyed himself in despair. Not being Catholics, we were ordered to inter them on the sea-beech. After we had buried them, the Native Creoles dug up the bodies of one

or

but of the moſt pleaſant appearance. The others are of great height, and may be ſeen at the diſtance of ſixteen or eighteen leagues. The Northernmoſt is diſtant from Cape

---

or two, and left them to be devoured by the dogs and vultures. On the ſame day the Spaniards ſailed with our veſſels, we were removed to Tepeak, a place ſixty miles up the country : here we were allowed great liberty, and better treatment; and more particularly ſo on the arrrival of Don Bodega Quadra, who was commander of his Catholic Majeſty's ſquadron, on the Coaſt of California. To this officer I am greatly indebted for his kind attention, and obtaining permiſſion for me to go to Mexico, to claim redreſs for our paſt treatment. On my arrival at Mexico, and during my reſidence there, I was treated by the Viceroy, Don Rivella Gigeda, with great politeneſs and humanity, and indeed by all ranks of people in that City. This Viceroy, in the moſt handſome manner, gave me an order to take poſſeſſion of my veſſel, and a paſſport directed to all claſſes of his ſubjects, to render me every ſervice I ſtood in need of whilſt in his government : and ſuch was his noble and generous treatment during my continuance of ſome months in Mexico and his ſubſequent correſpondence, that I am bound to acknowledge my laſting gratitude to him. I alſo underſtood the conduct of Martinez had, upon its being inveſtigated, occaſioned him very ſevere diſgrace. On my return to St. Blas, I found the Spaniards were unloading my veſſel, which had been laden with corn; and during my abſence, they had ſent her to Acapulco for guns and broke her back; ſhe was not only hogged, but otherwiſe greatly damaged, and they had alſo made uſe of every part of the ſtores, cargo and proviſions uſeful to them. For theſe they made out an account on a partial valuation of their own, and with an affected diſplay of liberality, calculated and allowed wages to my people, which they counterbalanced by charging them with maintenance, travelling expences, medical aſſiſtance, &c. &c. and alſo for an allowance of eight months ſtores and proviſions, *in which were included our beef and pork*, which we were obliged to ſalt before we put to ſea under a vertical ſun. After all, our departure was retarded, by their inſiſting I ſhould ſign a paper, expreſſing my complete and entire ſatisfaction of their uſage to me and my people.

Cape St. Lucas, which is the North Cape of the Gulf of California, fixty-five leagues; and the foutherumoft is diftant from Cape Corientes, which is the South Cape of

---

As the fever began again to make its appearance among fome of my crew, and the reft being extremely clamorous to depart, I was obliged, however, reluctantly, to fubmit. At length after thirteen months captivity, we obtained permiffion to fail, with orders to go to Nootka, and take poffeffion of the Princefs Royal, whofe crew I had with me, although the Spaniards muft have well known it was impoffible for me to have fallen in with her there, as appeared by the orders which the Spanifh commander had on board, when I met with him by accident fome time afterwards at the Sandwich Ifles. Thus on the approach of winter, in a miferable veffel, badly equipped, and worfe victualled, we failed from St. Blas, altogether in fuch a fituation, that from the numberlefs accidents we fuffered in confequence of our bad outfit, my arrival at Macao appeared almoft miraculous.

On my arrival at China, the refident agent D. Beal, Efq. who had taken no fmall degree of pains to inform himfelf of every particular concerning my capture, paid fuch of the crew as furvived the wages due to them, and requefted me once more to embark in the fame concern, on a voyage to Japan and Corea. I readily confented, and he fitted me out at a great expence, and in his inftructions to me, dated

Canton, July 25, 1791,

He fays — " After the mortifications and difappointments you have already experienced, from the capture of your veffels by the Spaniards, it may be an additional circumftance of regret, fhould difappointment and ill-fortune ftill purfue you: you muft, however, confole yourfelf, by reflecting that no imputation refts againft your character or conduct, for the violence and depredations committed by the Spaniards". This language from fo refpectable a character, was truly pleafing, and as an additional proof of his confidence, he fent his brother with me as fupercargo. But afterwards how great was my furprize, on hearing Mr. Mears had taken an advantage of my abfence, and publifhed in England an account of me fo contrary

to

of that Gulf, sixteen leagues. Between these isles, and about half-way to the main, are other small isles, called the Isabellas, which are remarkable for appearing, in all situations,

---

to truth! In his appendix to his voyage, speaking of the fever and delirium with which I was afflicted whilst in the hands of the Spaniards, he had stated on Mr. Duffin's ill-founded authority, that the delirium attending that fever was a family infirmity, and after wounding the feelings of all connected with me published the following by way of apology and reparation:

January 1, 1791.

"It is with particular satisfaction that I possess the opportunity of contradicting the mis-information of Mr. Duffin, relative to Captain Colnett's illness, in his letter to me from Nootka Sound, published in my memorial to the House of Commons, No. 9, and in the appendix to my voyages, No. 13: Mr. Duffin there mentions, but I am sure very innocently, that Mr. Colnett's insanity is supposed to be a family disorder; it therefore becomes my duty to declare, from the best authorities, that such a report is distant of any foundation whatever."

On my return to England in 1792, part of the money produced by the sale of the furs, mentioned in my introduction, was placed in one of the first banking houses in London, in the names of Messrs. Mears and Etches, to pay the amount ascertained at that time which was due to the heirs or assigns of such of the seamen as died on the voyage. If there are any monies remaining due to their representatives, &c. for loss of clothes and private property not yet settled, I am not accountable.

The most particular papers relating to the transaction at Nootka, being lost in his Majesty's frigate the Hussar which I had the honour to command in December, 1796, I have here given as circumstantial a detail as I can remember, from so long a period as nine years past.

situations, when at a small distance, like ships under sail. That part of the gulph, which lies between the Tres Marias and the main, forms a deep bay of fifteen or twenty leagues, and affords a good and safe anchorage, having regular soundings from the shore, and at the distance of four or five miles, five fathom; but whether the soundings extend to the Tres Marias, I have had no opportunity to inform myself: but when the Isabellas bore North, half East, distant five miles, I had good anchorage in twenty fathom water, muddy bottom.

The native Indians have a large establishment in this bay, known, in most of the charts, by the name of Mazatlan, but pronounced by the Creole Spaniards, Mauskelta town. It is remarkable for the great quantity of large fish, not unlike salmon in size and shape, which, during the summer season, are taken in the mouth of a small river near it: but previous to the capture of the vessels under my command, the inhabitants were unacquainted with a proper method of salting them. In this useful science they were instructed by some of my crew, who had been employed in the Newfoundland fisheries*.

Several

---

* The salting of this fish proved, however, a very unpleasant circumstance to us, as it occasioned our being employed to salt beef and pork for a fleet,

1793. Several other shallow rivers empty themselves into this bay, the principal of which is called Saint Jago, on whose Southern side, at the distance of two or three miles from the mouth, is situated the town of Saint Blas, that contains the grand arsenal and dock-yard of the province of Mexico, and is the chief depot for all the riches collected in the Californias. The principal store-houses and treasury are built on a small mount, that rises in the middle of the marsh which joins the dock-yard, and is about two miles from it. The face of the mount towards the sea is a perpendicular rock of one hundred fathom, and presents a very formidable appearance; but, on the land side, gradually sinks in several places to the plain. In the rainy season, when I was there, the marsh was so overflowed, as to render it a matter of difficulty to pass on foot to the dock-yard. There are not even at spring tides, more than ten or twelve feet water on the bar,

at

---

a fleet, then fitting out in the spring, at Saint Blas; with which the Spaniards were so well satisfied, that they took for themselves all the European salt provisions they found on board the vessels which they had captured; as having no doubt, but we could salt our own provisions when we should be released, which happened at the time the Sun was vertical; in consequence of which, though we did contrive, by cutting the meat in small pieces, to make it take the salt, yet, when we got out to sea, it was totally spoiled, and we were threatened with famine.

at the entrance of the river, and the frigates belonging to the ſtation in the Gulf of California, though they are capable of carrying fifty guns, are conſtructed ſo, as to paſs over the bar, and to protect the ſettlements on the gulf, from the attacks of the native Indians; who are continually at war with the Spaniards, particularly on the Eaſt ſide, which is ſaid to contain the richeſt mines of gold, that have been yet diſcovered; beſides ſeveral of ſilver.

The deepeſt water at the entrance of this river is cloſe along ſide the North point: where, on a gentle aſcent, there is an irregular battery of fourteen or fifteen pieces of cannon, of different bores, which they fetched from Acapulco, in one of my veſſels.

If I am correct in my recollection, for I have loſt all the minutes I made on the ſubject, it is high water on the bar of the river, at full and change, at ten o'clock, and the tide flows only eight or ten feet*.

When

---

* The ſhore in the bay is low; but the in-land mountains are very lofty; one of them which has the moſt ſingular appearance, is called Tepeak, and may be ſeen at the diſtance of thirty leagues. Here, myſelf and thoſe of my officers and crew who ſurvived the yellow fever at St. Blas, paſſed the ſix latter months of our captivity.

1793.

When the marine ſtores, &c. were brought by the way of Europe and Vera Cruz, a diſtance of eighteen hundred miles, on the backs of mules, Acapulco was the grand dock-yard; but ſince theſe ſupplies for the navy have been procured at a far cheaper rate by the way of China and Manilla, the naval arſenal has been removed to Saint Blas; before I left that place, the Viceroy of Mexico was ſo alarmed, leaſt the Court of Great-Britain ſhould revenge the inſult offered her by the capture of my veſſels, that, fearful of truſting to his flat-floored veſſels, &c. &c. he had ordered two heavy frigates of a ſharp conſtruction, to be built in the valley of Banderra, which is ſituated a few leagues to the Eaſt of Cape Corientes, for the better protection of the arſenal.

As I conceived it would be an act of the greateſt imprudence to anchor even near a Spaniſh port, I determined to return to the Iſle Socoro, in order to recover the health of the crew.

Nov. 12. We made the iſle on the twelfth day of November, and by the evening got well up with the North end; the Iſle Santo Berto being only eight leagues diſtance, and my not having had opportunity to aſcertain whether it afforded a better anchoring place than Socoro, determined me to examine it. We lay too all night for day-break, to make ſail, and by noon, got within three or four miles of the South end of

Santo

Santo Berto, when our Latitude by obfervation was 19° 15′, and Longitude corrected, 109° 54′. At this point, the ifland had a barren appearance, with little or no vegetation. It lay in a North Eaft, and South Weft direction, is about fix miles in length, and two or three in breadth, with a few rocks juft appearing above water off different parts of it. Its furface is uneven, and its appearance romantic; and, at the diftance of nine or ten miles, has the femblance of two feparate ifles. We faw fome feals there, and a great number of men-of-war hawks on the bluff, at the South end. On the Weft fide, is a fmall bay, but, as it difappointed my expectations, I did not land, or try for foundings in it. As the hurricane month and unfettled weather were not as yet over, and I knew of no fecure anchoring place at Socoro, where I could with fafety over-haul my rigging, and break up the hold, which we ftood in great need of, prepondering at the fame time in my mind, that the Ifle St. Thomas's did actually exift, and was not far diftant: I ftretched away to the Weftward in fearch of it, till we made 7° wefting, and reached the Longitude 118° Weft, in Latitude 20° 30′. I adminiftered to the crew who were afflicted with the fcurvy, twenty drops of elixir of vitriol, and half a pint of wine, three times a day, with fome preferved fruits, frefh bread, and pickles, from my own ftores, stores, and they began to mend. In our course, land birds frequently flew on board, particularly small grey owls, about the size of a black-bird; we were visited also by large horned owls, and brown hawks, as well as some of the size of our sparrow-hawks. They did not, however, come in such numbers as when we were off the Tres Marias and the Coast of California. From the above circumstances we were disposed to believe, we were in the vicinity of land: But I was more particularly encouraged in my hopes of seeing land, when, in Latitude 20° 25′, and Longitude 113° 27′ West, having fallen in with five or six wild ducks, the whaling master pursued them for some time in the boat; but, though they were not shy, he was not so fortunate as to kill one of them. Having joined the track of my former voyage in the Argonaut from St. Blas, which stretched 4° 30′ more to the Westward in the same Latitude, I gave up the idea of the island, which was the object of my immediate search, laying to the Westward of me; and not falling in with it on my return to Socoro, I cannot account for its situation, unless, according to the opinion of some modern hydrographers, it should be the Island Socoro itself.

On the twenty-fourth day of November, at day-light, we saw Rocka Partida, and passed to the Northward of it. At noon, on the same day, Socoro bore East by South, distant, seven or eight leagues.

1793.
Nov. 24.

On the following day, at noon, we got within a few miles of the South West end of that island: Latitude, by observation, 18° 49′ North. The boats were now hoisted out to search for an anchoring birth; and a small bay soon after appeared, which was formed by the South West and South terminations of the isle, wherein soundings were obtained, at twenty-five fathom, with a sandy bottom. We accordingly shortened sail, and came to anchor, at about the distance of two miles from the nearest part of the shore; the extremes of the island bearing from West North West to East South East; two small sandy beaches bearing from North by East, to North North East.

25.

On the twenty-sixth, A. M., I permitted the greatest part of the crew to go on shore, at a small cove, which was the only good landing place; and also put two men on shore abreast the ship, to look for water. In the evening, they all returned, with a considerable store of prickly pears beans and fish; the latter were of the snapper kind, and weighed

26.

1793.   weighed from four to eight pounds. Those of the crew who had perceived any symptoms of the scurvy laid themselves for some time, in the fresh earth, and derived considerable benefit from it: those who advanced up the country, saw many trees laying in a decayed state on the ground, which appeared to be of a much larger size than any that were standing; but they saw no spring or pool of fresh water, and were not encouraged to continue their search for it, as the surface of the ground was covered with a fine loose cinder, that rendered the walking over it laborious and difficult ; and it was the less necessary to undergo further fatigue, as we had plenty of water on board; and I was, at this time, in such a state of health, as rendered me incapable of attending upon any inland expedition. In the North East part of the island, where the ground was more firm, we afterwards found small quantities of water, lodged in the cavities of rocks; but, as that must have been supplied by showers, such resources must not only be insufficient, but uncertain. I have, however, no doubt, but that on the North East bay, wells might be made, that would produce plenty of good water; at least, the soil is such, as to encourage such an expectation: but a very heavy gale drove us to sea, before I was sufficiently recovered to make the experiment. The garden seeds which had been sown here, on our former visit,

were

were not come up, and the cocoa nuts, though they were in a growing ftate when we planted them, had decayed in the earth.

When we firft came too, off this bay, the wind was light to the Eaftward; but, at day-light, it blew ftrong from the NorthWeft, and Weft North Weft, and continued fo till eight in the evening of the twenty-feventh, when it became calm. During the whole of this day, the crew were fuffered to go on fhore; and, on its proving calm, we fhortened in the cable: but at midnight, by fome unaccountable accident, the anchor tripped; however, the fhip moft fortunately did not drive on fhore, if fhe had, would inevitably have been loft, as rocks extend for fome diftance off both points of the bay, and the light airs, which at intervals had blown, were moftly along the land. Not a perfon on board had the leaft fufpicion of what had happened till two o'clock in the morning.

It was a fingular circumftance, that having been reftlefs during the whole of the night, I quitted my bed at this hour, and went upon deck, when I mentioned to the officer of the watch, my fufpicion of the fhip's driving,

from

1793. from the found of the furf changing alternately on the points of the bay. I therefore ordered the deep fea-lead to be thrown overboard, and getting no bottom at forty fathom, my conjectures were inftantly confirmed. We now wore away fifty fathom of cable, but not bringing up, and a light breeze blowing, at the fame time, off the land, we backed off fhore, with the yards and mizen-top-fail. I can account for this accident in no other way, then from the too great length of the buoy rope, which, by the blowing of the variable light winds and the fhip's fwinging, had catched in her heel and weighed the anchor, which, with our crippled windlafs, employed us five hours to heave up.

Nov. 28. I now determined to have a tent pitched on fhore and land the fickly part of my crew, together with the fecond mate, who ftill continued to be in an infirm ftate, and beat off with the fhip, till they fhould be recovered. At noon, they were all got on fhore, and I left them the jolly-boat, to enable them to catch fifh; a diet at once both falutary and refrefhing to perfons in their fituation. In the afternoon, we ftood in with the North Eaft point, and kept the lead going, when we found regular foundings at five or fix miles, and from thirty-eight to ten fathom, at one mile and an half from fhore; at the fame time we were fheltered from the North Eaft,

East, to West by South. I now made a stretch off, bent my best bower, unstowed the other anchors, tacked and stood in, and came to in ten fathom water. The North East point bearing North, 45° East; the highest mount North, 33° 45′ West; the bottom of the bay North, 56° 15′ West; the Eastern point forming the entrance to the Cove, West; and the South point, West by South. In this situation we lay two days and a night, all hands on shore during the day, except one boat's crew: on the the third day, the current began to run to the North East, at the rate of two and an half, or three miles an hour, from which cause, we lay uneasy at single anchor. I was unwilling to moor with my bowers, as our windlass was in such a state as to render the heaving up an anchor a matter of great toil and delay; nor had we any boat to carry out a kedge sufficiently heavy to steady the ship.

Although the weather did not present the most promising appearance, and the winds Easterly, yet, as the current run to the windward, I entertained hopes of a long continuance of fine weather, which I always found at the Sandwich Isles, when the Northerly current run there. I was, however, mistaken; for in the night of the first

1793.
Dec. 1.

of December, the barometer fell fuddenly from 30-1 to 29-5-5, the winds hourly varied from Eaft to South, with fqualls, heavy fhowers of rain, continual lightning, and diftant thunder; which being on the approach of the new moon, fuch an alteration in the weather might be an expected event: but as the barometer had never deceived me, I was not fatisfied with its fudden change, and at the fame time entertaining doubts of the cable being injured, as the fhip had broke her fhear frequently during the night, I became very anxious for the dawn of day, to purchafe the anchor. At day-light, all round the horizon, and particularly from the South, threatened an inftant hurricane, which left me not a moment to hefitate for the fafety of the fhip, and with only eight hands on board, including myfelf, we rove a purchafe, weighed the anchor, and went to fea. As I conjectured, we found the cable fo rubbed and worn as obliged us to cut off twenty fathoms from it.

As foon as the fun had croffed the meridian, the heavy fqualls, and frequent fhowers of rain commenced, which continued to increafe till the change of the moon, at two o'clock in the morning, when it blew fo ftrong as to reduce us to clofe-reefed top-fails; and as the gale frefhened fo quick on us, we had not ftrength enough on board to fhorten any more fail, we were therefore obliged to carry it. We had

had now an heavy sea, torrents of rain, accompanied with thunder and lightning, and winds from every point of the compass, though principally from East to South East, which blew right into the roads we had left; and it is more than probable, from the state of our cable, and not laying more than a mile and an half from the shore, that, if we had attempted to ride out the gale, the ship would have been lost. It was, indeed, one of the worst nights I had experienced since I left Cape Horn.

On the third day of December, we got in with the shore again, and observing the jolly-boat alone, I felt the severest anxiety respecting the other boat and crew. We hove too, with the head off shore, and the whaling master was dispatched with every one on board, except myself, to ascertain what had become of them. Fortunately no accident had happened, except the wetting they had undergone from the violence of the rain, and the whale boat which I had missed, with some solicitude, had been taken by them on a fishing party, in order to bring a supply of fish on board the ship. I allowed the sick crew one day more to be on shore, and changed the party which was on board during the gale, to accompany them. During the whole night

night the weather was showery, with occasional lightning. The winds were well to the Eastward, and next day so much so, that I was obliged to carry a press of sail, to weather the North East points of the island, and could not therefore take the people from off the shore.

Dec. 4. On the fourth, at day-break, the winds inclining to the Northward, we run down off the cove, and got our tent and all hands on board by noon, anchors stowed, cables unbent, and made sail to the North East, for the Coast of Mexico, with the crew in perfect health, except the second mate; who, though he was much recovered, was still in a weakly and sickly condition. It may not be unnecessary for me to remark, that those of the crew who had any eruption on them of a scorbutic kind, I recommended them to bruise the prickly pear, and to apply the same in manner of a poultice, from which they not only found great relief, but it speedily recovered them, and much sooner then would have generally been credited.

Socoro, in the Spanish language, means supply; but during our stay at that island, we were not so fortunate as to discover any great affinity between the name, and character of the place. To this and the adjacent isles, I have given the name of Rivella Gigeda, after the viceroy of Mexico, as

the

the only return of gratitude as yet in my power, for the many acts of kindness and civility I received from him.

From a variety of observations of Sun, Moon, and Stars, I determine the Isle of Socoro to be in Latitude 18° 48′ North, Longitude 110° 10′ West, and bearing from Cape Corientes West, 22° South, distant ninety leagues. It lays in a West North West, and East South East direction; its greatest extent is eight leagues, and it is about three leagues in breadth. It may be said to consist of one mountain, which may be seen at the distance of twenty leagues, and falls in gradual descent at all points on the South side. It is in a great measure covered with brush-wood, intermixed with the low prickly pear-trees, and occasionally shaded with other trees of a larger growth. Some few spots of the soil are black and barren, as if fire had lately issued near it; and the top of the high land at a distance, has the appearance of there having been formerly a volcano: the surface is of a whitish colour, like that of the pumice stone, which was found on the shore. But though this may denote the existence of former eruption, I did not perceive either fire or smoke to issue from any part of the island. It must, however, be acknowledged, that Socoro is an excellent place of resort for a vessel with a scorbutic crew, or to refit if engaged in a cruize against the Spaniards off the Coast of Mexico, or employed in the whaling service.

The

The vegetables we found and considered as wholesome esculents, were beans and the molie tree, from whose leaves was made a very wholesome tea, of an aromatic smell and pleasant taste: but it is much smaller than that described by Mr. Falkner, though it was from his description of its leaf and fruit that I discovered it. The prickly pear, which is a very sovereign antiscorbutic, grows here also in great exuberance: it is of two kinds, white and red; but the former is considered as the most efficacious, and furnished us with the means of producing many wholesome, as well as palatable, pies and puddings. The animal food which we procured here, consisted of crows, owls, doves, black-birds, thrushes, sparrows, finches, and humming birds; besides water fowl — such as teal, sand larks, and various other sea birds, in great numbers. The fish we took were land-crabs, sea-crabs, craw-fish, colche with semicircular mouths, limpits, oysters, and other shell-fish*. To these may be added cod, rays, eels, and all those that are usually taken in tropical latitudes. The only novelty I found among the deep water fish, was one which bore some resemblance to the parrot fish, with a large hump of

---

* Of the species unknown before were the *Large Toothed Nevite*, the *Ribbed Green Turbo*, and the *Buccinum Dentex*.

of fat on the back part of its head. Of turtle, we saw only two, and caught neither of them. But with all this abundance of fish, it is a matter of some difficulty to obtain them, from the number and size of the sharks, who very frequently seized the whole of our prey, before we could draw them out of the water. Of quadrupeds, there were none visible to us: but of insects and reptiles, there were great numbers—such as spiders, flies, musquitos, grass-hoppers, crickets, and butterflies; with scorpions, lizards, and snakes. But the dearth of fresh water is the most uncomfortable and discouraging circumstance belonging to this island, though I am very much disposed to believe that an isle of this extent, and whose summit is continually covered with clouds, must have running streams on it: at the same time, the large flights of teal which are frequently seen coming from the interior parts of the island, strengthened my conjecture that it contains lakes pools, or springs, though it was not our good fortune to discover them.

The seasons of the year being considered, I think the safest anchorage from June to December is, between the South and South West points, opposite to two white coral beeches, which are the first two in succession from the South point of the island towards the West. It is

the

1795.

the place where we first anchored, and remarkable from the pinnacle rocks which lay close off the West point of the bay. I prefer this place in the bad season, as the wind seldom blows more than two points to the Southward of the East. In the good season, however, that is, from the latter end of December till the beginning of June, I prefer the South East bay, being better anchorage and nearer to the cove, which was the only good landing place we discovered, and is easily known, being a stony beech at the first inlet in the shore to the Eastward of the South point: all other part of the coast on the South side of the island is iron bound, which makes it extremely difficult, if not impossible to land, except in very fine weather.

According to the accounts given of the winds in this Latitude by former navigators, the South East bay would at all times afford a secure anchorage; but I found it otherwise: though such a change might be owing to the seasons falling later now than formerly, or in one year later than another. The Buccaneers assert, and Lord Anson confirms their assertion, that at the time he was cruizing for the Galeon, there was no reason to apprehend danger on the Coast of Mexico, from the middle of October till May. But my journal will shew, from
what

what we experienced, the beginning or middle of January is full early to expect good weather, for cruising, or fishing. To the Southward of Cape Corientes, and to the Northward of it Cape St. Lucas, the lightning, thunder and heavy rains had not subsided the beginning of November; and had not my crew been rather in a state of convalescence, I would have returned to the Northward for better weather. The Spaniards themselves never leave the Port of Saint Blas for Acapulco, till the latter end of November, when the North winds set in and blow steadily.

CHAPTER IX.

THE RATTLER QUITS THE ISLE OF SOCORO FOR THE COAST OF MEXICO: SOME ACCOUNT OF OUR TRANSACTIONS THERE, AND WHILE WE LAY AT ANCHOR BEFORE THE ISLAND OF QUIBO, IN THE GULF OF PANAMA, TO OUR ARRIVAL AT THE ISLES OF THE GALIPAGOES, ON AND NEAR THE EQUATOR.

1793.
December 6.
8.

IT was the fixth day of December, when we loft fight of Socoro; and on the eighth in the afternoon, we made Moro Corona on the Coaft of Mexico; we had pleafant weather and the winds were between the North Weft and the North Eaft. I entertained a ftrong defire to fee Paffion Ifle before I made the coaft, as it might have been of future advantage to fifhers and cruizers; but my bread was become fo bad as to be no longer in a ftate to be eaten, which made every perfon on board anxious to get

to

to the Southward and reach the Galipagoe Ifles where we might refit for England; unlefs we fhould fall in with fome European Veffel that would fupply us with the neceffaries which we fo much wanted; or from being made acquainted with the ftate of Europe, might venture into fome Spanifh port.

In our paffage to the Coaft, which we made in Latitude 19° 28′, we paffed great quantities of herring, turtle, porpoifes, black-fifh, devil-fifh, and fin-back whale, but the number of birds appeared to be greatly diminifhed fince we left the coaft: for at that time there were innumerable flocks of boobies, which were fo tame, as not only to perch on the different parts of the fhip, but even on our boats, and the oars while they were actually employed in rowing. When the appearance of the weather foretold a fquall, or on the approach of night, the turtle generally afforded a place of reft for one of thefe birds on his back; and though this curious perch was ufually an object of conteft, the turtle appears to be perfectly at eafe and unmoved on the occafion. The victorious bird generally eafed the turtle of the fucking fifh and maggots that adhere to and troubled him. We now faw dolphins and porpoifes in abundance, and took many of the latter, which we mixed with falt pork, and made excellent

faufages,

sausages, indeed they became our ordinary food. Sea snakes were also in great plenty, and many of the crew made a pleasant and nutritious meal of them.

We kept along the shore, under an easy sail, during the day and at night lay to. The winds were generally light and very variable, and we did not get off Acapulco till the nineteenth of December, the moon having then passed its full near three days, and the sun approaching to its greatest Southern declination. As we had not lately experienced any changeable or bad weather, we entertained the pleasing hopes that the unfavourable season was nearly passed, but at sun-set the blackest clouds I ever saw, gathered around us, and the succeeding night produced rain, with thunder, lightning, and heavy squalls of wind from all points of the compass, but chiefly from South to East. The rain continued to pour, in never-ceasing torrents, throughout the following day; but on the winds inclining to the North of East the rain began to abate, and towards the evening it fell only in heavy showers, and faint lightning continued to gleam through the night; but it was not till ten o'clock A. M. on the twenty-first, that the showers became moderate and we got sight of land: as we were within nine or ten leagues of it, with dark and unpromising weather, we made sail off shore with

with an Easterly wind; when, from the general bad state of my sails, I ordered the top-sails to be furled, and lay to under stay-sails. On the twenty-second of December the weather became moderate, with settled North Easterly winds and frequent showers, which continued without any variation to the end of this year. I shall not, however, omit to mention that, after the example of my first commander and patron, Captain Cooke, I did not suffer our Christmas, the grand festival of the christian world, to pass by without a sincere, though imperfect celebration of it.

We had now an alternate succession of calms and light winds, which blew from the North West quarter, and at times thunder and lightening. We proceeded down the coast under top-sails during the day, and lay to at night. When we saw any spouting fish, we stood off and on to ascertain their class, but of these there were very few, which proved to be hump-back and fin-back whale, black-fish and porpoises, but there were great numbers of albicores, bonnettas, dolphins and turtle, and of the two latter we caught as many as were necessary for our consumption.

On the thirty-first of December our Latitude was 14° 53′ and we had passed over the ground where we had reason to expect the greatest success in fishing, but had been driven off by

bad

1794. bad weather, without killing more than two or three whales; and as we did not now perceive the smallest trace of there being any fish of the spermaceti kind, and having every reason to believe, from the observations I had made, that their return like many other sea animals are periodical, under these doubtful circumstances it would have manifested an unpardonable degree of imprudence to have remained longer on this station with no more than six months provision, such as it was at two thirds allowance, and at such an immense distance from any of our own settlements. We continued for these reasons to pass under an easy sail along shore, flattering ourselves, at the same time, that we should either fall in with spermaceti whale, or meet with some vessel, who could afford us the assistance which we wanted. We now put the Rattler in the best posture of defence our situation would admit, as we were determined to speak to the first ship we met, and if she should prove an enemy, to trust either to our strength or superiority of sailing, the latter we had great faith in.

January 1. On January the first in Latitude 14° 36' we had a heavy gale of wind from the North East quarter, which occasioned a prodigious sea, and the ship to labour more than when she was off Cape Horn, so much so, that I was under some apprehension that we should lose our main mast. On the second

second the weather moderated, but became very changeable and foggy, with alternate calms and light winds. The night was moist with heavy dews, the colour of the sea frequently changed and there was much broken and white water. I kept the deep sea-lead constantly employed, but found no bottom at one hundred and fifty fathom, in Latitude 13° 33′ North. The winds westered on us and were succeeded by light and changeable breezes till we got into the Latitude 12° 48′, when we fell in with innumerable flights of those birds which are known to follow whale, and of which we had not seen such numbers since we were searching for the Isle Grande in the Atlantic Ocean.

1794.
January 2.

On the sixteenth we saw a sail to the Southward between us and the shore, and standing to the Northward and Westward. At noon, being in Latitude 12° 14′ 15″ North, we hove too to speak to her, our soundings were sixty fathoms, the volcano of Guatamala bearing North East by North, distant ten or eleven leagues. The vessel neared us considerably by one o'clock, and displayed Spanish colours: when it proved calm I sent the boat with the whaling master to board her, which he accordingly did, and returned with two sheep, six fowls, twelve tongues, several pumpkins and two bags of bread. The supercargo, who accompanied this present, brought an excuse from the master of the vessel,

16.

for

1794. for his making sail from us, which he attributed to the variable winds and his great anxiety to get to Acapulco, to which place he was bound from Lima. From this person I learned that Louis the Sixteenth King of France, had been beheaded by his own subjects, that the two Nations of Great Britain and France were engaged in war, and that there were on the Coast of Peru, a French privateer, two snows and a schooner, which had already captured several vessels. I sent the Spanish supercargo back to his ship, with a quantity of wine, rum, porter and cheese, which, far exceeded in value the present I had received, but it was impossible by any argument I could employ to procure any addition to it. The whaling master who was twice on board the Spanish vessel, might, on the first visit have had his boat filled with whatever he had demanded; but on his second appearance, the Spanish Commander had recovered his spirits but lost his liberality, for he would not part with any thing more. From his general conversation, and the manner in which he stated the probability of our being taken by the French cruizers as we went down the coast, we had some reason to believe that Great Britain was at war with Spain as well as France.

We soon parted company with the Spanish trader, and stood to the South, distancing the land, at the same time,

from

from twelve to fifteen leagues. The sea was continually varying in its colour, but we could not obtain any soundings.

On the twenty-third of January at noon, our Latitude was 8° 49′ 51″ North, Cape Blanco bearing North 3° East. Our stock of water was now very much reduced, and the greater part of that which remained, was, from its having been kept in oily casks, become so nauseous as to produce sickness instead of allaying thirst: I therefore made sail for the Island of Quibo, in order to obtain a fresh supply of such a material article, on which our future health depended. Our winds since we lost sight of Guatimala, were between the South East, and North East; and would at times vary for a few hours to the Western Quarter.

On the twenty-sixth we had moderate breezes from North West to South West, our Latitude was 7° 54′ North. On the twenty-seventh, being in the vicinity of the Isle Mentuosa, between Cape Dulce and Quibo, we fell in with several spermaceti whales, of which we killed four, and afterwards were so unfortunate as to lose one along-side. The sight of these whales prolonged our cruize until the eighth of February, in the hope of getting more of them, but we only added four to those we had already taken. The winds

1794. on this cruize were very variable, but rather more in the weſtern than the eaſtern quarter.

Between Cape Dulce and the South end of Quibo, are the Iſles Zedzones, Mentuoſa and Quicaras. The Zedzones conſiſt of ſmall barren rocks. Mentuoſa riſes to a conſiderable height, and is five or ſix miles in circumference, its ſummit is covered with trees, the greater part are thoſe which bear the cocoa nut, which gives it a very pleaſant appearance, but iſlets and breakers extend off its Eaſt and Weſt ends to the diſtance of three or four miles. The bottom is rocky on the South ſide, as is the ſhore near the ſea. There is a beach of ſand behind ſome little creeks that runs in between the rocks, which makes a ſafe landing for boats. Here we went on ſhore, and got a quantity of cocoa nuts with a few birds. The Spaniards or Indians had been lately here, to fiſh on the reef for pearls, and had left great heaps of oyſter ſhells. It may not, therefore, be improper to ſuggeſt to thoſe who may hereafter find it convenient to land in this iſland, to be prepared to defend themſelves, in caſe they ſhould be attacked by any of its occaſional viſitors. There were a great plenty of parrots, doves and guanos, and it is probable that other refreſhments might be obtained of which we are ignorant. At all events, it may

may be useful to whalers or cruizers, by offering a place where there sick may be landed, and cocoa nuts procured, whose milk will supply the want of water. This island, according to my observations, lies in Latitude 7° 15′ North, and Longitude 82° 40′ West. The quicaras consist of two isles: the larger one is about six or seven miles, and the lesser about two or three miles, in length; they lay North and South of each other, with but a small space between them; and distant from the South end of Quibo, about twelve miles. The least of these isles is entirely covered with cocoa trees; and the larger one bears an equal appearance of leafy verdure, but very few of the trees which produce it are of the cocoa kind.

The whole of my ship's company longed so much to get some good water to their bad bread, and our success in fishing had fallen so short of our expectations, that I was induced to quit the whaling sooner, than I should otherwise have done: therefore on the eighth day of February at Noon, we rounded the South end of Quibo, the Latitude by observation being 7° 19′ 25″ North, soundings thirty-eight fathom. The South point of Quibo bearing South 42° West, the North East point bearing North 45° West, and Cape Mariato bearing East 4° 30′ South. We had light airs and pleasant weather

weather, during the greater part of the afternoon, the winds were at South East by East, and we steered North, North West with all sail set to get to an anchor before night, keeping the lead constantly going, and during a run of eleven miles, our soundings were from thirty to thirty-six fathoms, and on drawing near to the North East point of Quibo, shoaled quick to ten fathom and an half, in which bottom we came to anchor; the North end of Quibo bearing North West by North; and the South end, South East by South. The boats were immediately sent to discover the wateringplace.

It was calm through the night and the early part of the morning, when we weighed anchor on the flood tide, to tow to a more convenient situation, but finding the water shoal to four fathom, and the bottom very visible, it was discovered that we were nearly surrounded by a reef which extended four or five miles from the shore. By the active conduct of the boats crew an anchor was carried out, and we warped off into ten fathom; a breeze then springing up from the East, we made sail, and ran along the edge of the reef, sounding seven, eight, nine and ten fathoms, at the distance of a mile and half from the shore. We soon after came to anchor and moored in the bay of Port de Dames in nineteen fathoms: the North point of the bay in a line with the North point of Isle Sebacco, bore North North East, the watering place North 44° West; and South point Isle Quibo South 32° East. Latitude by observation 7° 27′, and Longitude 82° 10′.

We

We lay here till the seventeenth of February, and got on board forty-three tons of water, with some fire wood. But of other refreshments we obtained little, though we had parties constantly employed in trying both the water and the land for fresh provisions. After all, two or three monkies, and a few doves, were all we got from the island; and its surrounding water afforded us only alligators, crabs, cockles, clams, periwinkles, oysters and a few other shell fish u iknown to us*. Several deer were seen among the thickets on the shore, as well as wolves, and the feet of some animals, which were supposed to be tygers, had left their impression on the sands. But the animals, were all of them so shy, that they kept beyond the reach of our fire-arms, and it was equally difficult to take the turtle which were seen in great abundance. That the birds and monkies were quickly alarmed, may be readily accounted for, from the numbers of hawks and large vultures who feed upon them; as in the maws of some of the latter which we killed, young monkies were found. The wolves and tygers may be supposed to keep the less offensive quadrupeds in a similar state of agitation; and the fish, as well as the turtle, may be harrassed into an equal alarm by the alligators, sharks, sea-snakes, &c. all of which, particularly the first of them, seem to swarm on and about the surrounding shores.

1794.
February 17.

From

---

* Viz. The green Trochus, the black Buccinum, Buccinum Morus Patula, and Subula, together with the Strombo, Tuberen, Latus and Patalla, not before well known to collectors in conchology.

1794.   From one of them I had a very fortunate efcape. As I was walking along the fea coaft, with a gun, and very attentive to the woods, in expectation of feeing fome kind of fowl or game proceed from the thickets, fuddenly my danger was difcovered, of having paffed over a large alligator, laying afleep under a ledge of the rock, and appeared to be a part of it; and being in a deep hollow I could not have efcaped, if a little boy, the nephew of Captain Marfhall, who accompanied me, had not alarmed me with his out-cry. I had juft time enough to put a ball in my gun, the noife having roufed the hideous animal, and he was in the act of fpringing at me when I difcharged my peice at him, its contents entering befide his eye, and lodging in his brain, inftantly killed him; it was then taken on board, where part of him was eaten. In the ftomachs of feveral of the fnakes which we took, there were fifh in an undigefted ftate, and of a fize that credulity itfelf would almoft refufe to believe. Thefe voracious animals, appear to have greatly leffened the quantity of fifh on the fhores of this ifland, which afforded fuch an abundant fupply of delicious and falutary food to former navigators. The woods alfo abound with fnakes of different kinds, the largeft we faw were the hooded fnakes. As I was fetting on a bank at the fide of a rivulet, one of the fmaller bit me by the left knee, which

caufed

caufed it to fwell to that degree, that I had a doubt for fome time whether it would not coft me my life.

The vegetables and fruits we obtained on this ifland were but few. There were fome cocoa trees in the bottom of the bay; and we found beans growing near the fpot, where the Spanifh pearl fifhers or Indians had refided; and from whence, as we conjectured from the ftate of their fire-places, they were but lately removed. The miftol and the chanmer tree, mentioned by Mr. Falkner, were feen in great plenty, but the fruit produced a naufea and ficknefs foon after it was fwallowed. The officer, whom I fent to the Northward, informed me, that the huts remained which are mentioned in the voyage of Lord Anfon, and confidered that bay as the moft convenient for any fhip that might be obliged to remain at this ifland to refit.

Quibo is the moft commodious place for cruizers, of any I had feen in thefe feas; as all parts of it furnifh plenty of wood and water. The rivulet from whence we collected our ftock, was about twelve feet in breadth, and we might have got timber for any purpofe for which it could have been wanted. There are trees of the cedar kind a fufficient fize to form mafts for a fhip of the firft rate, and of the quality which the

Spaniards

Spaniards in their dock yards use for every purpose of ship building, making masts, &c. A vessel may lay so near the shore as to haul off its water; but the time of anchoring must be considered, as the flats run off a long way, and it is possible to be deceived in the distance. The high water, by my calculation is at half past three o'clock; at full and change the flood comes from the North and returns the same way, flowing seven hours and ebbing five, and the perpendicular rise or the tide two fathoms. I found several betel nuts which appeared to have been washed on the shore by the tide, but I did not see any of the plants that bear them, growing on the shore, though several of my people, after we had left the place, mentioned their having seen many of them.

It would not be adviseable for men of war and armed vessels, acting upon the defensive or offensive, to anchor far in, as the wind throughout the day, blows fresh from the Eastward, and right on shore, so that an enemy would have a very great advantage over ships in such a situation. There is good anchorage throughout the bay; at five or six miles distance, thirty-three and thirty-five fathom, with a mud bottom, and firm holding ground.

The most commanding look out is the top of Quicara, we saw it over Quibo (which is low and flat) while we lay

lay at anchor; and is, I prefume, the remarkable mountain which Lord Anfon miftook for part of Quibo as mentioned in his voyage. Indeed, a good look out on the top of this ifland may be neceffary for many obvious reafons, as it commands the whole coaft and bay. We intended going to fea the feventeenth at day-light, but the difficulty we had in purchafing the anchor from the good quality of the bottom, delayed us until the fea breeze fet in, fo that we could not fail till the eighteenth. We faw while here one fail, and fhe was fteering to the South, between Quibo and the main. On leaving Quibo, we cruized between the Ifle Quicara, and Cape Mariatto, till the laft day of February; during which time, we killed feven whales; fix of which we got along fide, and loft one by breaking a drift in the night. We afterwards faw another, but it was fo blafted as to be of no ufe. As the Sun now drew near the equator, and long calms were to be expected, it became neceffary for us to reach the Galipagoe Ifles before they commenced; where we propofed, (as the whaling bufinefs had failed,) to procure falt, for the purpofe of falting feal-fkins at the Iflands of Saint Felix, and Saint Ambrofe, in Latitude 26° 15′ South.

1794. The different navigators of thefe feas have given fuch various accounts of the paffage from hence to the Galipagoes, that it became a matter of fome perplexity, to determine which route to be preferred. While we were cruifing between the South end of Quibo and Cape Mariatto, the winds were light and moftly Southerly. They fometimes blew a ftrong gale through the night, but generally a ftiff breeze from North by Eaft, to North by Weft: but in the day we had pleafant weather. As I could depend on the failing of the Rattler, I determined

March 1: on my route the firft of March, and fteered away to the Southward in a direct line for the ifles.

4. On the fourth day of the fame month, being in Latitude 4° North, the winds varied between the South Eaft and South Weft points, and at intervals blew from the Weftward; but when they returned to the Northward, they were very light and of fhort duration. At this period an innumerable flight of birds accompanied us, and we had turtles in great plenty, but they foon grew fcarce; though we continued to take bonnettas, dolphins, porpoifes and black-fifh in great abundance. The weather then changed to rain with thunder and lightning; and we every day remarked our

paffing

paſſing through ſtrong ripplings and veins of currents, all of which run to the Weſt till we made the iſles.

1794.

On the twelfth, at break of day, we ſaw Chatham Iſle, and, by ſun-ſet came to an anchor in Stephen's bay, near the South Weſt point of the iſle in twenty-eight fathom water; the two points of the bay bearing North Eaſt and South Weſt, and the Kicker rock, bearing Weſt, North Weſt, at the diſtance of two miles. We attempted to get into this bay to the Weſtward of the rock, but as there was little wind, with a current running right out, and no foundings to be got, with fifty fathom of line, till within three quarters of a mile of the ſhore, and then a rocky bottom, we hauled out to the North, and went in to the Eaſtward of the Kicker rock, there being regular foundings between it and the bluff, which formed the Eaſtern point of the bay: the greateſt depth between them thirty fathoms, but the deepeſt water is near the rock.

March 17.

We lay in this bay till the ſeventeenth of March, employed in ſearching for ſalt, procuring a ſtock of turtles, and recovering ſeveral of the crew, who were afflicted with boils, they were ſoon reſtored by the fruit

17.

of the molie tree, wild mint tea, and a diet of turtle and teal soup, &c. Our boats traversed all the lee-side of the isle for salt, but without any success; though they discovered several rills of fresh water. One of them proceeded from a bluff which forms the East point of the bay, and others were seen at the bluff at the Eastern part of the isle. The latter were not examined, as the party did not land there; and the former was no more than sufficient to fill a ten gallon cag in a quarter of an hour. As these high bluffs are at the extremity of the low land, the rills must proceed from some bason or lake on the interior high grounds. One of these I afterwards found on a hill which I ascended, from whence the water was entirely drained. On the coast of America, in the dry season, I have seen a long succession of lagoons of this kind, without the smallest drain on the beach below. The head of Stephen's bay possesses the convenience of a small interior cove, with three fathom water, that will hold four or five sail, and where they would be sheltered from all winds. Also a fine sandy beach beneath the rocks, on which a vessel may be hauled on shore, or heave down if occasion should require it; and great abundance of turtles, mullet, and other fish might be caught in a seine. The turtles pass over the rocks, at high water, into salt lagoons to feed. The land is so low in this part of the island, as,

at

at a small distance, to give it the appearance of being divided by a channel of the sea. Near the West part of the isle in a small bay was a part of the wreck of a ship, that appeared to have been but lately cast away, as a whole wale plank was found undecayed. On some of the small isles in this bay, were the largest prickly pear-trees I had ever seen.

After weighing from Stephen's bay, it was with great difficulty we cleared it by night, from the light, variable winds and torrents of rain. When we had got well out, we hove to for day-light, and then made sail for an isle which bore from our anchoring birth, West by South, to West by North. By noon of the next day, we saw many more isles and islets to the North and Westward of us: and at sun-set, we saw breakers a long way to the Northward and Westward of Lord Hood's isle. Our Latitude at Noon was 0° 31′ 51″ South. We now shortened sail and stood on and off for the night. The next day we found ourselves set considerably to the Southward and Westward; and in sight of Charles Isle, so named by the Buccaneers. At noon our Latitude was 1° 28′ 13″ South; the extremes of Charles Isle bearing from West 6° North, to West 29° North. In the early part of the evening we got close in with the South end of the island: we then shortened sail, and stood off and on

during

1794.

March 20.

1794.

during the night, with the defign of going on fhore in the morning. This ifle is of a moderate height, prefents a pleafant afpect, and is furrounded with fmall iflets, the two largeft of which I named after the admirals Sir Alan Gardner and Caldwell. There are feveral fandy beaches on it, and a great number of feals were feen off it. At day-light the current had fet us fo confiderably to the Southward and Weftward, as to have loft fight of the ifland, though we plyed to Windward all the forenoon we gained but little. We got fight, however, of Albemarle Ifle, and two fmaller ones which lie between it and Charles Ifle. I take them to be the Croffman and Brattles Ifles of the Buccaneers.

March 20. At noon on the twentieth, our Latitude was 1° 23′ South: the extremes of Charles Ifle bearing from Eaft 14° North, to Eaft 24° North; and Albemarle ifle from North 45° Weft, to North 10° Weft; with a fmall flat ifle between them. We faw feveral fpermaceti whales, and gave chafe with boats and fhip but could not come up with them. We beat off here for forty hours, and loft ground confiderably from the current running fo ftrong to the Weftward.

21. At noon on the twenty-firft, our Latitude was 1° 19′ South, Albemarle Ifle bearing from North 20° Eaft, to North 31° Weft; and Perry Ifthmus, North 5° Weft. By four o'clock in the afternoon, we got within two miles of the South

and

and East end of Albemarle Isle, when we tried for sounding with one hundred fathom of line but found no bottom. The following day, as soon as it was light, we bore up to round the South and West end of Albemarle Isle, called, by the Buccaneers, Christopher's Point. Within a few miles of it, the Latitude was, by observation, 0° 55′ 14″ South. The extremities of Albemarle Isle, bearing from East 22° South, to North 10° East; and of Narborough Isle from North, to North 20° West.

1794

A large bay opened to our view, which was formed by the South and West points of Albemarle Isle, and the East part of Narborough Isle, having received originally from the Buccaneers the name of Elizabeth Bay. As it is very capacious, we conjectured that we should find good anchorage; I therefore accompanied the chief mate to examine it, but we could find no bottom for two leagues at the distance of a mile or a mile and an half from the shore, with one hundred and fifty fathom of line. The inhospitable appearance of this place was such as I had never before seen, nor had I ever beheld such wild clusters of hillocks, in such strange irregular shapes and forms, as the shore presented, except on the fields of ice near the South Pole. The base appeared to be one entire clinker to a considerable distance

March 23,

from

from the water-fide, and the little verdure that was vifible was on the tops of the hills, which were crowned with low, fhaggy bufhes, that gradually diminifhed in quantity as they hung down the declivities; and were fometimes divided by veins of an hard, black, fhining earth, which, at a fmall diftance, had the appearance of ftreamlets of water. The ftorm peterels accompanied us in great numbers: but the wind coming right out with a current or tide, that was fo rapid, as to be attended with fome degree of danger, we gave up our defign of reaching the head of the bay, particularly as night was approaching, and darknefs would have overtaken us. When I returned on board, I found the fhip laying between two winds, and becalmed within half a mile of the fhore, where no bottom could be obtained with one hundred and fifty fathom of line. In this fituation we were near an hour, with flaws of wind all round the compafs, and heavy fhowers. At laft, we caught a Southerly wind and made fail to the Weftward, and when clear of the fhore, hove to for the night. The weather was dark and gloomy, with heavy dews and a ftrong foutherly current; fo that at day-light we were fet nearly as far to the South as we were on the preceding noon. At noon our Latitude was 0° 35′ 6″ South: the extremities of land bearing from North 12° Eaft, to Eaft 37° South.

In

In the evening we got well up with the South end of Narborough Isle, and stood along to the North Westward, by the West shore. The current or tide had now changed its course, and set, from the West and South, to the Northward, directly on that isle, and the night proving calm, with some difficulty we cleared it; for we could not find any bottom at the distance of half a mile from the shore, with one hundred and fifty fathom of line. At the return of day the weather was dark and cloudy, with lightning in the South East. At noon I observed on the Equator, the extreme points of Narborough Isle, bearing from South 21° East, to South 52° East. The North West Cape of Albemarle Isle, (which I have named Cape Berkeley, from the honourable Captain Berkeley), bearing East 4° North, North end East 27° North. The North point of land in sight, bearing East 36° North, and the Rodondo Rock North 5° East, at the distance of five or six leagues.

I sent away a boat in the forenoon to sound a large bay, formed by the North end of Narborough Isle and Berkeley point, (which I have named Banks's Bay in honour of Sir Joseph Banks), or under Berkeley point, in order to discover a place of anchorage: the boat, however, did not get

1794. get into the bay; but rowed under the North point of Albemarle Ifle, where the party landed, and returned in the evening. They found this part of the Ifle equally inhofpitable as the Southern part of it: but had procured a few rock-cod, with fome hump-back turtles, and faw a confiderable number of feals.

Narborough Ifle is the higheft land among the Galipagoe Iflands, lying near the center of Albemarle Ifle, which almoft furrounds it, in the form of two crefcents, and making two bays. The apparent point of divifion of thefe iflands, is fo low on both, that I am in doubt whether they are feperated. On the next morning we faw fpermaceti whales, we killed feven and got them along fide; Rock Rodondo bearing Eaft 22° South, the Northernmoft land bearing Eaft 18° South, and the South Weft land bearing South 28° Eaft. The weather was hazy, and the Latitude by obfervation

April 8. 00° 27′ 13″ North. Here we cruifed till the eighth of April, and faw fpermaceti whales in great numbers, but only killed five, of which we fecured four. The current ran fo ftrong to the Weftward, and the winds were fo light, that after laying to, to fecure the whales and cut them up, we were feven days in returning to the ground from whence

we

we drifted. In the winter feafon, when the winds are more frefh, thefe difficulties might not occur, otherwife, it would be impoffible for any veffel, which was not a very prime failer, to whale here with fuccefs; though at a certain feafon any quantity of fperm oil might be procured. The oldeft whale-fifhers, with whom I have converfed, as well as thofe on board my fhip, uniformly declared that they had never feen fpermaceti whales in a ftate of copulation, or fquid their principal food in fhoals before; but both thefe objects were very common off thefe ifles, and we frequently killed the latter, of four or five feet in length, with the granes. Young fpermaceti whales were alfo feen in great numbers, which were not larger than a fmall porpoife. I am difpofed to believe that we were now at the general rendezvous of the fpermaceti whales from the coafts of Mexico, Peru, and the Gulf of Panama, who come here to calve: as among thofe we killed, there was but one bull-whale. The fituation I recommend to all cruizers, is between the South end of Narborough Ifle and the Rock Rodondo: though great care muft be taken, not to go to the North of the latter; for there the current fets at the rate of four and five miles an hour due North. Narborough Ifle falls gradually down to a point at the North, South, and Eaft ends, and may be equal in produce to any of the neighbouring ifles; but of this I can only

con-

1794. conjecture, as I did not myself examine it; nor does it appear that the Buccaneers ever landed upon it.

The Rodondo is an high barren rock, about a quarter of a mile in circumference, and is visible as far as eight or nine leagues, has soundings round it at the distance of a quarter of a mile thirty fathom. Here our boats caught rock-cod in great abundance. I frequently observed the whales leave these isles and go to the Westward, and in a few days, return with augmented numbers. I have also seen the whales coming, as it were, from the main, and passing along from the dawn of day to night, in one extended line, as if they were in haste to reach the Galipagoes. It is very much to be regretted that these isles have to this period been so little known but only to the Spaniards.

Though we met with so strong a current, it did not dishearten us, as we found, by keeping between the North point of Narborough Isle, and North point of Albemarle Isle, and not going to the Northward of the latter, that we were able to maintain our ground; and the hope which now possessed us of making a very successful voyage, dispersed every complaint of bad bread and short allowance, which were no longer considered either with regret or impatience.

We

We recovered the fishing ground after having been driven off during four days, and found as great plenty of whales as when we left it. We now saw a ship in shore, who sailed well, and was heavy mettled as we conjectured from the report of a gun. I discovered with the telescope that she was French built, and from the intelligence communicated by the Spaniard we fell in with off the Gulf of Guatamala, on the Coast of Mexico, we had every reason to believe that she was one of the French ships which he mentioned as being in these seas. We kept standing in with the shore to reconnoitre her, having great confidence in the sailing of our own vessel. During the evening, night and morning, we had alternately heavy fogs, light winds and calms. At nine A. M. the weather became clear. I now stood towards the sail, but the nearer I approached the more I suspected her to be an enemy. I then stretched away to the Southward, when she carried every thing after us, and getting a strong Northerly breeze, which she brought up with her, over-reached us very fast. We made all the sail we could from her, (our Latitude at noon 0° 19′ 52″ North,) but I entertained little or no hope of escaping: we therefore cut down the stern, in order to get out two three-pounders, which were all the great guns we had, and put

ourselves

1794.
April 8.

9:

1794. ourselves in the best posture of defence in our power. Finding at four o'clock in the afternoon that she still gained ground upon us, but would not be able to get up with us till it was dark, we all agreed to a man, to heave to, and if she proved an enemy, to board her; as such a desperate proceeding would be altogether unexpected, we thought it would afford some of us a better chance of escaping, than by a more regular engagement. As to myself, death, in almost any shape would have been far preferable than falling again into the hands of the Spaniards. By sun-set, however the ship joined us, and proved, after all our alarm and preparations, to be the Butterworth of London, Mr. Sharp, from a trading voyage on the North West Coast of America, and lately from California. We were right in our conjectures concerning her appearance, as she was taken from the French in the last war. She had been searching for water in these isles but had found none; and was bound to the Marquises for it, with only seven butts on board; a route of near eight hundred leagues, when there were so many places within two days sail, where she might have found it. Mr. Sharp had sixty tons of salt in bulk, for the purpose of salting skins; and on the coast of California, he had procured an hundred tons of oil from the sea lion and sea elephant; and he added, that he also might have procured

ten

ten thoufand tons of oil from the fame animals, if he had poffeffed a fufficient number of cafks to have contained it.

I recommended him to proceed to James's Ifle, and offered him a copy of a chart, which I had received from Mr. Stephens, which would direct him to the watering place, defcribed by the Buccaneers, whofe information I had no reafon to doubt: but if he had no faith in it, he might go to Ifle Cocas or Quibo, where I had procured plenty; but no perfuafion of mine, however, had any weight, as his principal object appeared to be that I fhould accompany him. In addition to my other inclinations to render him every fervice in my power, the feveral acts of civility I had received from Mr. Perry of Blackwall, one of his owners, had the greateft weight with me; and underftanding his intention was alfo to continue in company to our arrival in England, I undertook to fhew him the way into port.

In confequence of light winds, thick weather and ftrong Northerly currents, we were driven as far North as 1° 5′, and faw Culpepper's Ifle, which rifes to a confiderable height, though it is of fmall extent; but the weather was

fo

so hazy, and we were at such a distance, that I am not qualified to give a further account of it.

Though our ships were excellent sailers, we were fifteen days in getting into James's Bay; they alternately had the advantage of each other; but the Rattler was entirely out of trim, the fore-hold being filled with oil. The Butterworth had so far got the advantage to windward, as, at one time, to be within a few miles of the anchoring ground; and we could only see her top-gallant sails; she bore up to join us again, with only three butts of water on board. At this time we were close under Abington Isle, which is very small, and was well known to the Buccaneers; and, according to my observation, is in Latitude 0° 33′ North, and Longitude 90° 45′. It is high towards the South end, which has a very pleasant appearance, and where is the only bay or anchoring place in the island. The North end is low, barren, and one entire clinker, with breakers stretching out to a considerable distance. I sent a party in the boat to round it, where they caught plenty of small fish with their hook and line. They also landed on the island and found both tortoises and turtles. This day we also saw Bindloes Isle, which is a small, rugged spot, laying to the Southward and Eastward of Abington Isle, and about the mid-way between it and James's Isle.

On

On the twenty-fourth, in the very early part of the afternoon, we came to an anchor at the North end of James's Isle, a little to the South of Fresh-water bay, where the Butterworth followed us; Albany Isle bearing North 34° West; bottom of the bay East 17° South; South point of James's Isle, on with Cowley's enchanted Isle, and South part of Albemarle Isle South 24° West: North point of Albemarle Isle West 25° North.

1794.
April 24.

As soon as the ship was secured, I set out with Mr. Sharp to search for water in Fresh-water bay, where the Buccaneers had formerly supplied themselves, but the surf prevented us from landing. We rowed close to the beach, but saw not the least signs of any spring or rivulet. Boats were dispatched from both the vessels to different parts of the shore; and my chief mate was sent away to the South for a night and a day. On the following morning at dawn of day, the whaling-master was ordered to land if the surf was fallen, and search Fresh-water bay. He accomplished getting on shore, but found no water; and in the evening, the chief mate returned with the same account of his unsuccessful errand. For my own part, I never gave up my opinion that there was plenty of water in the isle; but as neither of my boats were in a condition to encounter the least

U

bad

bad weather, I deferred taking a furvey of the ifle till they were repaired.

Though we fent the Butterworth daily fupplies of water, I did not forefee the confequence of our generofity; for from that moment, the commander never gave himfelf the leaft concern to look for any; but employed his crew in cutting a very large quantity of wood, and ftocking himfelf with land tortoife privately, from a fpot which we agreed fhould remain facred, till we were ready for failing, and then fhare our ftock together. Indeed I not only fupplied Mr. Sharp with water, but may be faid alfo to have added to his food; for he did not know that the tortoife was an wholefome eatable till I informed him of it.

As I had at this time many reafons to doubt his continuing long in company with me, and in cafe of feparation the Rattler had no boat belonging to her calculated to bring water any diftance, it awakened my precaution to provide for any unforefeen accident fhould it befall us refpecting that neceffary article. I determined therefore, to fupply him monthly throughout our voyage, and the information of this arrangement produced a better effect than I expected, as it ftimulated him to fearch for water, which he found within two miles of his fhip.

After

After anchoring and his prefent wants being accommodated, he varied fo in his future plans, to his former ones propofed, that I could not comprehend he had any fixed one at all; and his conduct in general not correfponding to my ideas or expectations, I had only to lament, that after putting myfelf to fo great an inconvenience, there was fo little probability that it would be attended with any advantage to his employers. Finding my advice of no farther ufe I failed without him.

1794.

As foon as a boat was repaired, I fet out to furvey the South Eaft part of this and Albemarle Ifle. On reaching the South point of James's Ifle, I got fight of three other ifles which I had not feen before, nor can I trace them in the Buccaneers accounts, no more than the ifle which we faw to Weftward, when at anchor in Stephens's bay, Chatham Ifle. Thefe three ifles now feen, I named after the admirals Barrington, Duncan, and Jarvis. The two Northernmoft, which are neareft to James's Ifle, are the higheft, and prefented the moft agreeable appearance, being covered with trees. The Southernmoft, which I named Barrington Ifle, is the largeft and was the greateft diftance from me, it is of a moderate height, and rifes in hummocks; the South end is low, running on

a parallel

1794. a parallel with the water's edge. We did not land on either of them. In this expedition we saw great numbers of penguins, and three or four hundred seals. There were also small birds, with a red breast, such as I have seen at the New Hebrides; and others resembling the Java sparrow, in shape and size, but of a black plumage; the male was the darkest, and had a very delightful note. At every place where we landed on the Western side, we might have walked for miles, through long grass and beneath groves of trees. It only wanted a stream to compose a very charming landscape. This isle appears to have been a favourite resort of the Buccaneers, as we not only found seats, which had been made by them of earth and stone, but a considerable number of broken jars scattered about, and some entirely whole, in which the Peruvian wine and liquors of that country are preserved. We also found some old daggers, nails and other implements. This place is, in every respect, calculated for refreshment or relief for crews after a long and tedious voyage, as it abounds with wood, and good anchorage, for any number of ships, and sheltered from all winds by Albemarle Isle. The watering-place of the Buccaneers was entirely dried up, and there was only found a small rivulet between two hills running into the sea; the Northernmost of the hill forms the South point of Fresh-
water

water bay. Though there is a great plenty of wood, that which is near the fhore, is not large enough for any purpofe, but to ufe as fire-wood. In the mountains the trees may be of a larger fize, as they grow to the fummit of them. I do not think that the watering-place which we faw, is the only one on the ifland; and I have no doubt, if wells were dug any where beneath the hills, that it would be found in great plenty : they muft be made, however, at fome diftance from the fandy beach, as within a few yards behind them, is a large lagoon of falt water, from three to eight feet in depth, which rifes and falls with the tide; and in a few hours a channel might be cut into it. The woods abound with tortoifes, doves, and guanas, and the lagoons with teal. The earth produces wild mint, forrel, and a plant refembling the cloth-tree of Otaheite and the Sandwich Ifles, whofe leaves are an excellent fubftitute for the China tea, and was indeed preferred to it by my people as well as myfelf. There are many other kinds of trees, particularly the moli-tree, mentioned by Mr. Falkner, and the algarrooa, but that which abounds, in a fuperior degree, is the cotton tree. There is great plenty of every kind of fifh that inhabit the tropical Latitudes; mullet, devil-fifh, and green turtle were in great abundance. But all the luxuries of the fea, yielded to that which the ifland afforded us in the land tortoife,

which

which in whatever way it was dressed, was considered by all of us as the most delicious food we had ever tasted. The fat of these animals when melted down, was equal to fresh butter; those which weighed from thirty to forty pounds, were the best, and yielded two quarts of fat: some of the largest, when standing on their feet, measured near a yard from the lower part of the neck. As they advance in age their shell becomes proportionably thin, and I have seen them in such a state, that a pebble would shatter them. I salted several of the middle size, with some of the eggs, which are quite round, and as big as those of a goose, and brought them to England. The most extraordinary animal in this island is the sea guana, which, indeed abounds in all these isles. We did not see the land guana in any of the isles but James's, and it differs from that which I have seen on the coast of Guinea, in having a kind of comb on the back of its neck.

These isles deserve the attention of the British navigators beyond any unsettled situation: but the preference must be given to James's Isle; as it is the only one we found sufficient fresh water at to supply a small ship. But Chatham Isle being one of the Southernmost, I recommend to be the first made, in order to ascertain the ships true situation fituation, in which you may be otherwife miftaken, from the uncertain and ftrong currents, as well as the thick weather which is fo prevalent there. As it ftands by itfelf there is no danger, and in Stephens's bay, thirty or forty fail may ride in fafety, befides thofe which might go into the cove. Veffels bound round Cape Horn to any part North of the Equator, or whalers on their voyage to the North or South Pacific Ocean, or the Gulf of Panama, will find thefe iflands very convenient places for refitting and refrefhment. They would alfo in future ferve as a place of rendezvous for Britifh fifhing fhips, as they are contiguous to the beft fifhing grounds.

CHAP.

CHAPTER X.

THE RATTLER LEAVES THE GALIPAGOE ISLES AND COAST OF PERU, FOR THE ISLES SAINT FELIX AND SAINT AMBROSE, ON THE COAST OF CHILI: FROM THENCE SHE ROUNDS CAPE HORN, ON HER PASSAGE TO ISLE SAINT HELENA, IN THE ATLANTIC OCEAN.

1794.
May 13.

ON the thirteenth of May, having over-hauled the rigging, caulked, wooded, &c. we fet fail with the intention to cruize for feven days off Rock Rodondo, and then to proceed to the Ifles Saint Felix and Saint Ambrofe, on the coaft of Chili. We accordingly hove to for the night, off the North end of Albemarle Ifle, and at break of day, faw feveral fpermaceti whales, of which we killed two. The winds had fet in from the Southward and Eaftward, with a ftrong Northerly current; fo that all our endeavours were in vain to get to the Weftward and round to the South,

South, without wasting as much time as we had before done, to get to the Eastward, when we wanted to reach James's Isle. From the South the current set from three to four miles an hour, due North, and we had in general, thick, foggy weather. We frequently saw whales; and on the 16th of May, got sight of Wenam's Isle, bearing West North West, seven or eight Leagues. It is small, but of considerable height, like Culpepper's Isle, and I make it in Latitue 1° 21' North, and Longitude 91° 46' West. The time of our proposed cruize off these isles was expired, and the winds obliged me to stand away to the Eastward and Northward, with the strong current setting against me, to the Westward and Northward; so that I was fifteen days making Cape Blanco, the South Cape of the Gulf of Guiaquil, a distance we had run before in four days. Half way over we fell in with a body of spermaceti whales, we got up with them, though not without some difficulty, and killed three, but were so unfortunate as to have two boats stove in the struggle.

1794.

May 16.

Within Cape Blanco, we saw a sail crouding every thing from us, which induced us to conjecture that it was no longer peace between Great Britain and Spain. But this vessel was too far up the Gulf, as well as in

1794.
June 5.

too shallow water for us to follow her. On the following morning, being the fifth of June, we got a steady wind from the South West, but as we distanced the shore and Southerd our Latitude, it hauled to the South East, encreasing daily in strength, with an heavy sea. The weather was sometimes squally, with frequent showers of rain; and when we got into Latitude 17° South, and Longitude 90° West, the wind hauled well to the East.

19.

On the 19th of June, when we were in Latitude 24°, and Longitude 90° 30′, an heavy gale of wind blew from the Northward. From the time of our leaving Cape Blanco the ship had made water, which now began to gain

21.

on us: and in the afternoon of the twenty-first, in a violent squall of wind and rain, our fair weather top-sails and courses were blown to pieces, and having neither canvas or twine to repair them, we were under the necessity of bending our best and only suit.

At night, being in the supposed situation of Saint Felix and Saint Ambrose Isles by different navigators, we hove to till day-light, and then scudded till night and again hove to, as we did, on the succeeding night, at which time the weather moderated. Having now run down both to the East

East and West in the supposed Latitude of these isles, I am convinced that there are no other near this situation than those I visited in my outward-bound passage; and where I was at this time determined to land a party for the purpose of salting and drying seal-skins; intending then to proceed to the Eastward as far as Easter Isle, to search for isles mentioned in the following extract of a letter in the possession of Philip Stephens, Esq., and of whose existence I entertain not the least doubt, as in their description they differ much from Easter Isle, which I visited with Captain Cook, there not being a tree on it.

EXTRACT.
16th September, 1773.

" The Achilles left Calloa the seventh of April, and arrived at Cadiz the tenth, by which we learn that the frigate Le Lievre (the Hare) had discovered five islands in the South sea, in about 27° of South Latitude; that one of them was considerably large, and inhabited by Indians somewhat tractable, and governed by a chief. They have hatchets and other utensils, which they say the English left there three months before the Lievre arrived there."

EXTRACT.
27th September, 1773.

" The tenth instant came into Cadiz, the merchant ship Achilles, which left Calloa off Lima, the seventh of

1794. April laft, This veffel brought news to the court of the difcovery, and the taking poffeffion, in the name of the King, of feveral fmall iflands in the South feas, to 27° of Latitude South of Lima. There is one ifland rather large, and has an excellent harbour. This ifland they have called Saint Charles; and the difcovery was made by the King's frigate the Eagle, which the Viceroy of Peru fent upon that expedition. They fay that thefe iflands are inhabited by favage Indians, but that they were very well difpofed; and that the country abounded with wood, fowls, hogs, and certain roots of which they made bread, perhaps caffada."

"It would appear that the court means to make ufe of this difcovery; and that they mean to build forts thereon, and to eftablifh a communication between thefe iflands, and the continent of South America. It is probable that the famous Mr. Hudfon had difcovered thefe ifles in his voyage round the world, and that the largeft of them are called Davis's land. What renders this conjecture more probable is, that they found the interior inhabitants poffeffed of hatchets, fpades and hoes."

Roggewein's account of an ifle in this Latitude, differs fo much from Eafter Ifle, that I cannot fuppofe it to be the fame. Mr. Wafer, who was furgeon with Captain Davis, in 1685, and after whom land in this Latitude is named, differs very widely from Roggewein's account, and alfo Captain Cook.

1794.

If I had not found thefe ifles, the potatoes which I entertained the hopes of procuring at Eafter Ifle, would have enabled me to lengthen my voyage, and to double Cape Horn in the fummer feafon.

On the twenty-fecond at noon, we made the Ifles Saint Ambrofe and Saint Felix, and prepared every thing for landing. During the laft twenty-four hours, the wind had hauled to the Southward, and we had to beat up againft it. Throughout the night it blew very ftrong in fqualls, while the fhip laboured very much, and the leak encreafed fo as to keep both pumps employed. By the quantity of water perceived in the hold, we fufpected that it rufhed in forward, and that part of the wooding ends were ftarted.

June 22.

On

1794.　On the North fide of the Wefternmoft ifle, at half a mile diftance from the fhore, there was fafe anchorage, with a foutherly wind, which now blew: but as we had fo lately experienced an heavy Northerly gale, which is the prevailing wind in winter, and blows directly into the anchoring birth, the general opinion was to make fail back to the Northward, to get into better weather or in with the main land, and endeavour to ftop the leak.—In fhort, any fituation however inconvenient, or even dangerous, was preferred by the whole crew, to the putting into a Spanifh port, and trufting to the tender mercies we might find there. It becomes an act of juftice in me to declare that, in every awkward and unpleafant circumftance, in which we fometimes found ourfelves, every perfon on board, from the whaling-mafter to the loweft feamean, manifefted a perfect confidence in me, and paid an implicit obedience to my opinion.—But the fuperftition of a feaman's mind is not eafily fubdued, and it was with fome difficulty that I could preferve an hen who had been hatched and bred on board, and who at this time was accompanied by a fmall brood of chickens, from being deftroyed, in order to quit the ill omen that had been occafioned by the unexpected crowing of the animal during the preceeding night.

On

On the twenty-third, Latitude 26° 0', the weather moderated so much as to afford an opportunity of examining the leak, when we found the lower cheek of the head loosened, and the wash-boards of the starboard cheek, entirely washed away; the oakum worked out of the wooden ends, so as to admit an arm-full to be stuffed in by hand, and no one was yet convinced but that the plank had started from the stem. We made our utmost exertions to get every thing aft, in order to raise the leak above water: and here, to add to our disappointment, it became necessary, for want of food to sustain them, to kill our small stock of pigs which had been reserved to regale us on our homeward passage round Cape Horn.

1794.
June 23.

By the twenty-seventh we had returned again to the Northward as far as 18° South, when we finished caulking and leading over the leak, the only method we had of securing it, having neither pitch, tar, or rosin on board, our marine stores being all expended. Our bread was not fit to eat, and our other provisions so short, that owing to its bad qualities we could scarce exist on it: thus situated, we proposed as our last trial in these seas, to continue on to the Northward till we made the land, in hopes to fall in with

27.

with some European vessel to obtain supplies to enable us to wait for a more favourable period to round Cape Horn.

How far I may be right in my conjecture must be decided by future trials, but I am very much disposed to believe, that the tar we had on board was of a bad quality, and destroyed not only the copper but iron, and was, in some degree, the cause of our leak: for the copper, wherever it was paid with it, was become as thin as paper, and the copper-headed nails, as well as those of iron, had received considerable injury.

June 29. On the twenty-ninth we reached as far Northward as
30. 16° 50′ South, and made the coast of Peru; on the thirtieth at noon we were within a few miles of the shore, and not seeing any ships, we conceived our opinion of a war with Spain was confirmed, and the only rational alternative left us, was to brave all the difficulties that we experienced and was further threatened with; and force ourselves as soon as possible out of them, by losing no time in getting round the Cape into the Atlantic; this being determined we took our departure for England. As we stretched to the Southward, the wind hung more to the Eastward of South, than on the former part of our voyage.

When

When we were in Latitude 24°, a very singular circumstance happened, which as it spread some alarm among my people, and awakened their superstitious apprehensions, I shall beg leave to mention. About eight o'clock in the evening an animal rose along-side the ship, and uttered such shrieks and tones of lamentation so like those produced by the female human voice, when expressing the deepest distress, as to occasion no small degree of alarm among those who first heard it. These cries continued for upwards of three hours, and seemed to encrease as the ship sailed from it: I conjectured it to be a female seal that had lost its cub, or a cub that had lost its dam; but I never heard any noise whatever that approached so near those sounds which proceed from the organs of utterance in the human species. The crew considered this as another evil omen, and the difficulties of our situation were sufficient, without the additional inconvenience of these accidental events, to cause any temporary depression of those spirits which were so necessary to meet the distresses we might be obliged to encounter.

As we sailed up the coast of Chili and Peru, from the Latitude 38° South, we never had occasion to reef from the strength of the wind; while the barometer, from that

1794. Latitude, stood mostly at 29-9, and the thermometer at 60, rising gradually till in the Latitude of 1° 30′ South, till it reached 72; but in the evening, it was generally below summer heat in England. Along the whole of this coast, the dews were very heavy during the night: and in proportion as they were heavier, the succeeding day was more or less clear. At the full and change of the moon we perceived no dew, which appeared to be supplied by an heavy drizzling rain and misty weather. The morning, evening, and night, were always cloudy, but the middle of the day was generally clear, so that I seldom enjoyed a distinct view of the Cordileras des Andes. The mistiness of the early part of the day, proceeded from the sun rising behind the Andes, and the clearness of the noon was occasioned by the sun, which had then over-topped the mountains; but I am yet to learn the cause of the haziness of the evening.

The currents on this coast are very irregular. I tried them several times, and found that they set as often one way as the other, and generally from half a mile to two miles an hour. The set, may at all times be discovered by observing the direction of large beds of small blubber, with which this coast abounds, and from whence the water

derives

derives a colour like that of blood; I have sometimes been engaged for an whole day in passing through the various sets of them.

The fish, common to this coast, are dolphins, and all those which inhabit tropical Latitudes; and in calm nights, there are seen large shoals of small fish which have the appearance of breakers. Of turtle, we saw none till we were North of Lima, they were of that kind called the loggerhead, and North of the Equator we found the hump-backed species on the surface of the water in great numbers. We frequently took out of the seals and porpoises large quantities of squid, which is the food of the spermaceti whales, and at times we saw many devil-fish and sun-fish, the latter of which proved an agreeable and wholesome addition to our daily fare.

All the birds which are usually seen at sea in similar Latitudes are to be found on this coast. There are also the Port Egmont hen and albatross, which are generally supposed to be the constant inhabitants of colder climates.

We sometimes passed great numbers of small birds, lying dead on the water; a circumstance for which I am not able to assign a probable conjecture. The greatest number of pelicans appeared off Lobas le Mar, and if that place should be their constant resort, they will, in thick weather, determine the vicinity of the island.

I tried for soundings, in many parts of the coast, at the distance of five and six leagues from the shore, but could not obtain any bottom with one hundred and fifty fathoms of line. In thick weather, however, when you draw near land, large quantities of sea-weed will appear, and birds, in great numbers, sitting on the water. Seals are no certain criterion for being near the shore; as I have often seen them, at the distance of an hundred and fifty leagues from land, sleeping in great numbers on the surface of the water, with the tail and one fin out of it, so as to offer the appearance of a crooked billet. On any part of the coast of Chili, or Peru, a sealing voyage might be made with great prospect of success, as well as at the Isles of Saint Felix and Saint Ambrose.

In our passage down the coast of Chili, we had South East and Easterly winds, with variable, but in general

pleasant

pleasant weather, accompanied with occasional showers. In Latitude 33° South, the wind Southerd on us and the next day veered to the West, and continued mostly between the West and North till we got into 47° South. It would sometimes blow, for a few hours, between the West and South West, but never continued. In the Latitudes of 48° and 49°, the winds were light for forty-eight hours in the South East quarter, with a strong Southerly current.

On the twenty-sixth of July, in Latitude 48° South, the coast of Chili presented to us a range of high mountains covered with snow. We had now frequent showers of rain, hail and snow, and, on the first of August, doubled Cape Horn at the distance of fifteen Leagues. During the whole of the passage, the weather was not, by many degrees, so bad as we had apprehended, and was much better than that we had experienced when we came from Europe.

When we had rounded the Cape, and had advanced to the North, the weather improved every hour. In the Latitude 49°, the wind blew for twenty-four hours in the South East quarter, with delightful weather. Our spirits as may be supposed, were greatly cheared by such a

favourable

favourable paſſage, and were in a ſtate to be enlivened by the ſea-birds who flew twittering around us.

During the ſucceeding twenty-four hours, the winds varied from North Weſt to North Eaſt, and became at laſt very changeable. The paſt hour we were hurried along by a ſtrong gale, and the next at reſt in a dead calm. At noon our Latitude was 47° 30′, Longitude 48° 40′, with a very heavy irregular ſea, in which the ſhip greatly laboured: This laſted, however, but for a few moments, when an heavy gale from the South Weſt ſprang up, which was accompanied with rain, hail and ſnow.

Under reefed fore-ſail, and cloſe-reefed main-top-ſail, all the ſail we could carry, we ſhaped our courſe, on the neareſt angle, to Saint Helena, but before midnight, the ſea roſe to a prodigious height, broke on board of us, and ſtove in the dead lights, filled the after part of the ſhip with water, rendered uſeleſs a chronometer, a ſextant, and deſtroyed charts and drawings that I had been ſeven months employed in completing: alſo damaged every thing in the cabin. We ſoon, however, fixed and ſecured temporary dead lights, and pumped out the water, but ſome of the miſchief done was irreparable.

When

When we were at our greatest Southern Latitude, the thermometer stood at 42-5, and the barometer was never lower than 28-8-0. In the last gale, the thermometer stood at 38-5, and barometer 28-7-6, which was the lowest point to which it sunk during the voyage.

Between the Latitude 53° and 40° South, and Longitude 59° and 38° West, we saw large bodies of seaweed, and great numbers of birds: and on the eleventh of August, we crossed near the supposed situation of the Isle Grande. At this time my vessel was almost a wreck, very short of provisions, and what remained in a very bad state, to which may be added an hurricane of wind and the winter season: circumstances that, I trust, will be a sufficient excuse for my not renewing my search of it as I had intended.

The wind remained in the South West quarter, during five days, at which period our Latitude was 35° 45′, and Longitude 31° 22′ West, when we had light and variable winds. On the eighteenth of August, at noon, the Latitude being 33° 41′, the wind settled in the North East quarter, and blew a fresh breeze for four days, but on the succeeding four, it varied round the compass, with frequent rain. By the

1794.
Auguſt 31.

the laſt day of Auguſt, in Latitude 19°, the wind inclined to, and continued in, the Eaſt and South Eaſt quarter.

Sept. 1.

On the firſt of September, at Noon, we made the Iſland of Saint Helena, after a paſſage of one month from Cape Horn. At this time I had no more than two of my crew, who were afflicted by the Scurvy, and the ſame number beginning to complain, which was not ſo much owing to the length of the Voyage as to their own want of care and cleanlineſs after getting out of the South Seas and never ſhifting their wet clothes. The diſeaſe ſeized them in a manner very different from any appearance of this diſorder which I had yet ſeen: they were principally affected in their hips firſt, and then down their legs. We had one man indeed, who was literally panic-ſtruck by the appearance and cries of the ſeal in the Pacific Ocean; if we had remained twenty-four hours at ſea, he would not have recovered.

CHAPTER XI.

FROM ISLE SAINT HELENA TO ENGLAND.

———

1794.
Sept. 2.

AT ten in the morning of the fecond of September, we anchored in James's Bay, Saint Helena, and found riding there, an outward-bound Eaft-Indiaman, and an American brig, from the Cape of Good Hope to Bofton. I waited on the Governor who received me with great politenefs, and gave me a general invitation to his houfe. The fame civility I alfo received from the Lieutenant Governor, and was offered a fupply of every thing I wanted from the Company's Stores. I now became acquainted with the war between Great Britain and France; but it was very uncertain when a convoy would arrive, I determined therefore, as my veffel was a very fine failer, to make my ftay here as fhort as poffible; and accordingly,

1794.
Sept. 13. by the thirteenth, the Governor having made up his packet, we sailed for England in perfect health.

23. On the twenty-third of September, being in Latitude 4° 38′ 9″ North, and Longitude 23° 22′ West, the wind
28. varied to the Westward; and on the twenty-eighth, in Latitude 24° 22′ North, and Longitude 24° 3′ West, it got to the Northward of West, and continued to be variable between the North East and North, North West to South West. From this time we had very changeable, squally
October 11. and thick weather till we made land. On the eleventh of October, the head of our mizen-mast was gone; and on
15. the fifteenth, in a squall, the head of the-main-mast sprung. On approaching the Western isles we housed the boats, knocked down the try works, and fresh painted the ship, in order to assume as much as possible the appearance of a man of war. We saw several sail, between this arrangement and our making land, but did our utmost to avoid them.

Nov. 1. On the first of November, we made the Eddystone Light-house, and after reaching as high as we could, we hove to Dartmouth and sent the letters on shore. In the course of the night we reached Portland; and stood off and on on for day-light, when we ran up and anchored in Cowes road, Iſle of Wight.

This voyage occupied twenty-two months, and after doubling Cape Horn we met only with one Engliſh and two Spaniſh ſhips in the Pacific Ocean; nor did we touch at any known port but the Rio Janeiro in going out, and Saint Helena on our return home. It is not the leaſt of my ſatisfactions to mention, that except the loſs of one man by an unforeſeen accident, the whole of the crew conſiſting only of twenty-five men and boys, were preſerved during this long, fatiguing and perilous voyage.

FINIS.

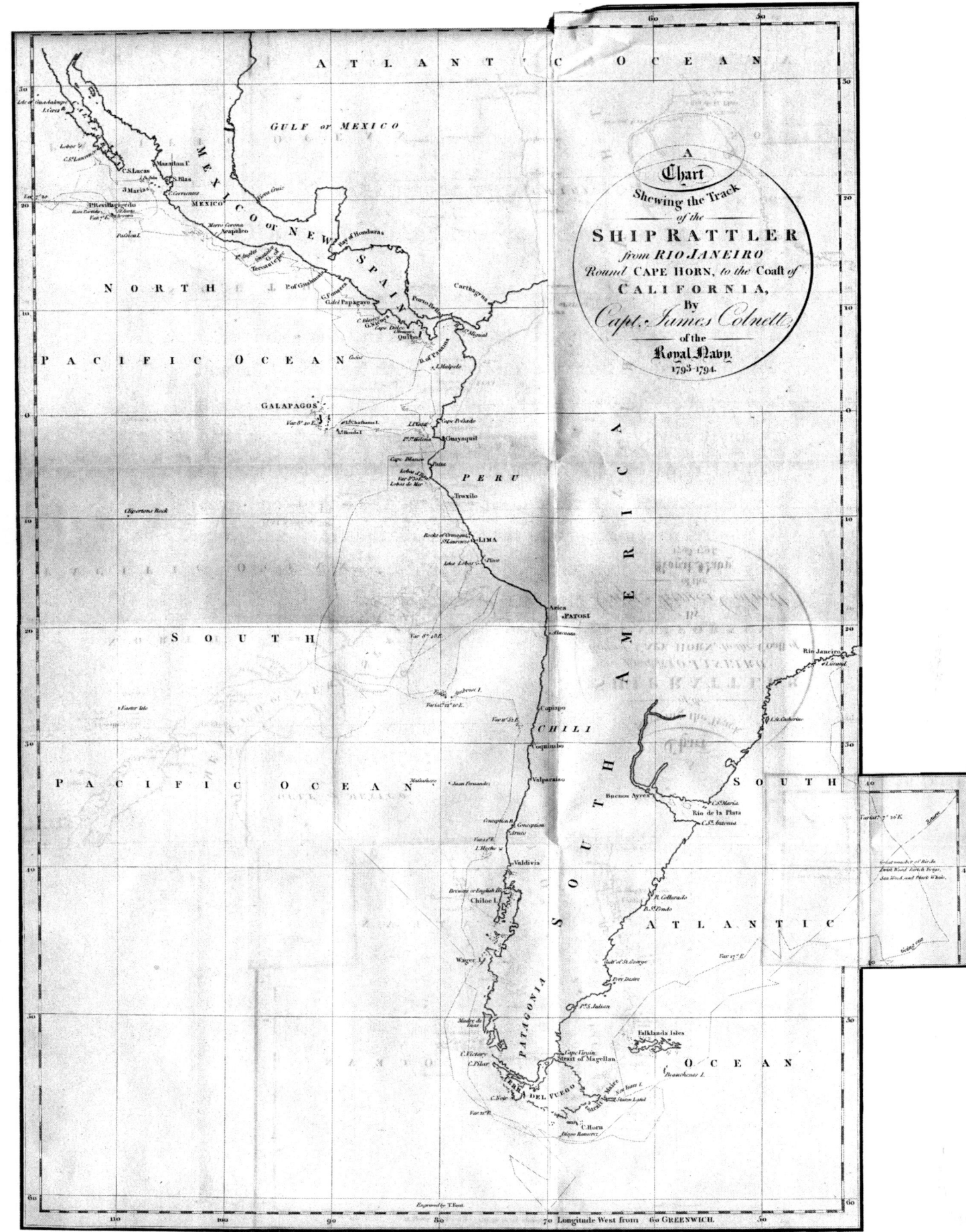

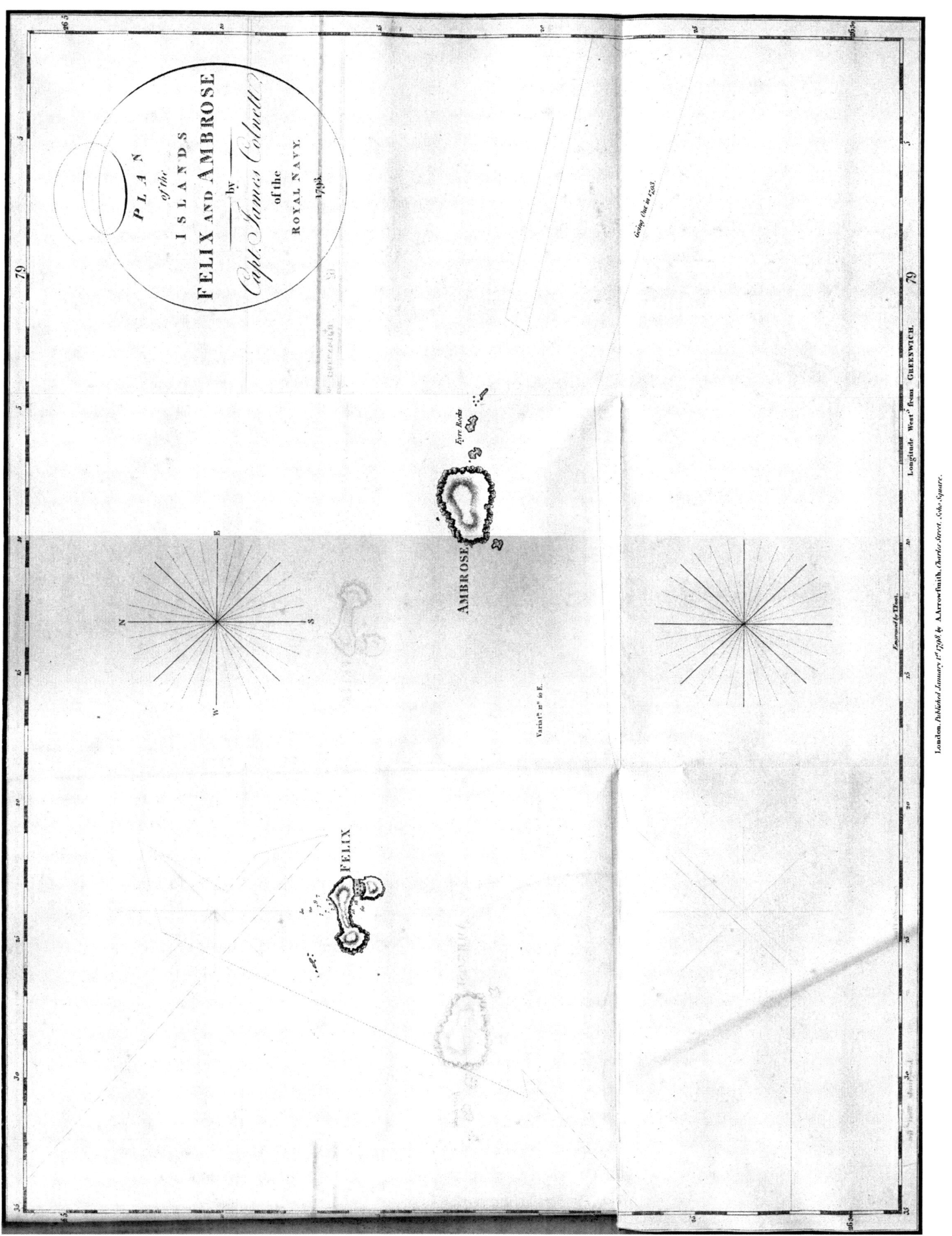

The material originally positioned here is too large for reproduction in this reissue. A PDF can be downloaded from the web address given on page iv of this book, by clicking on 'Resources Available'.

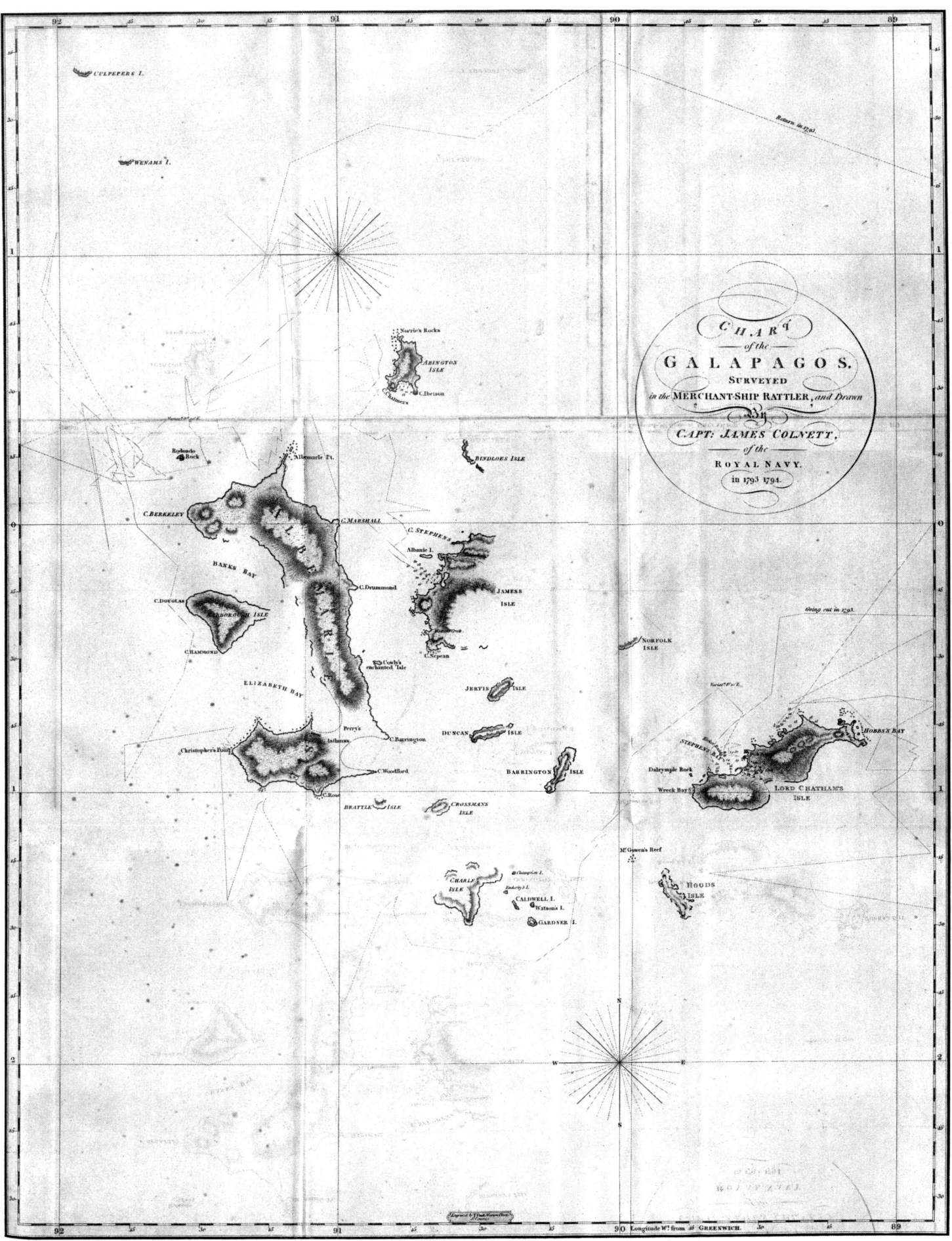

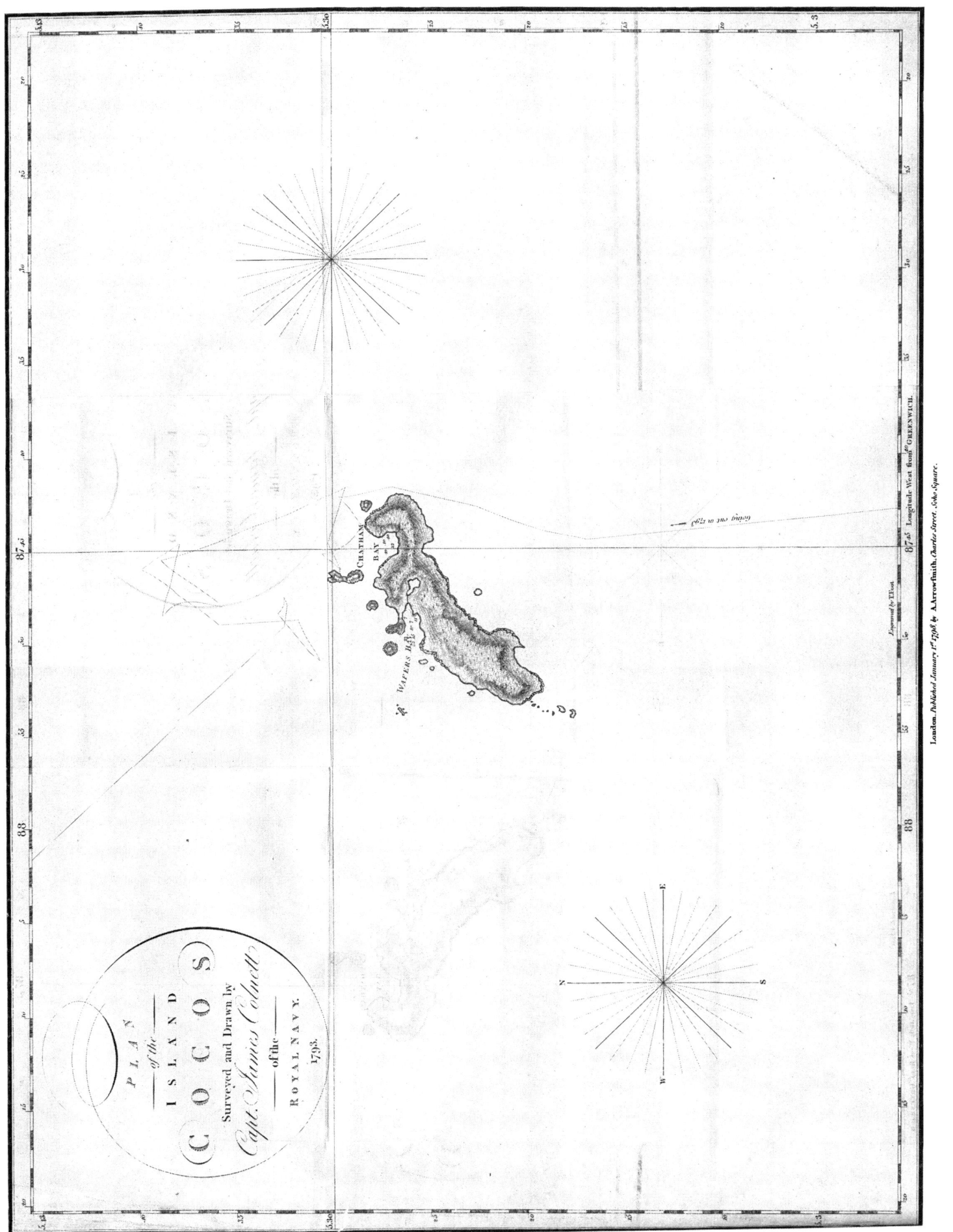

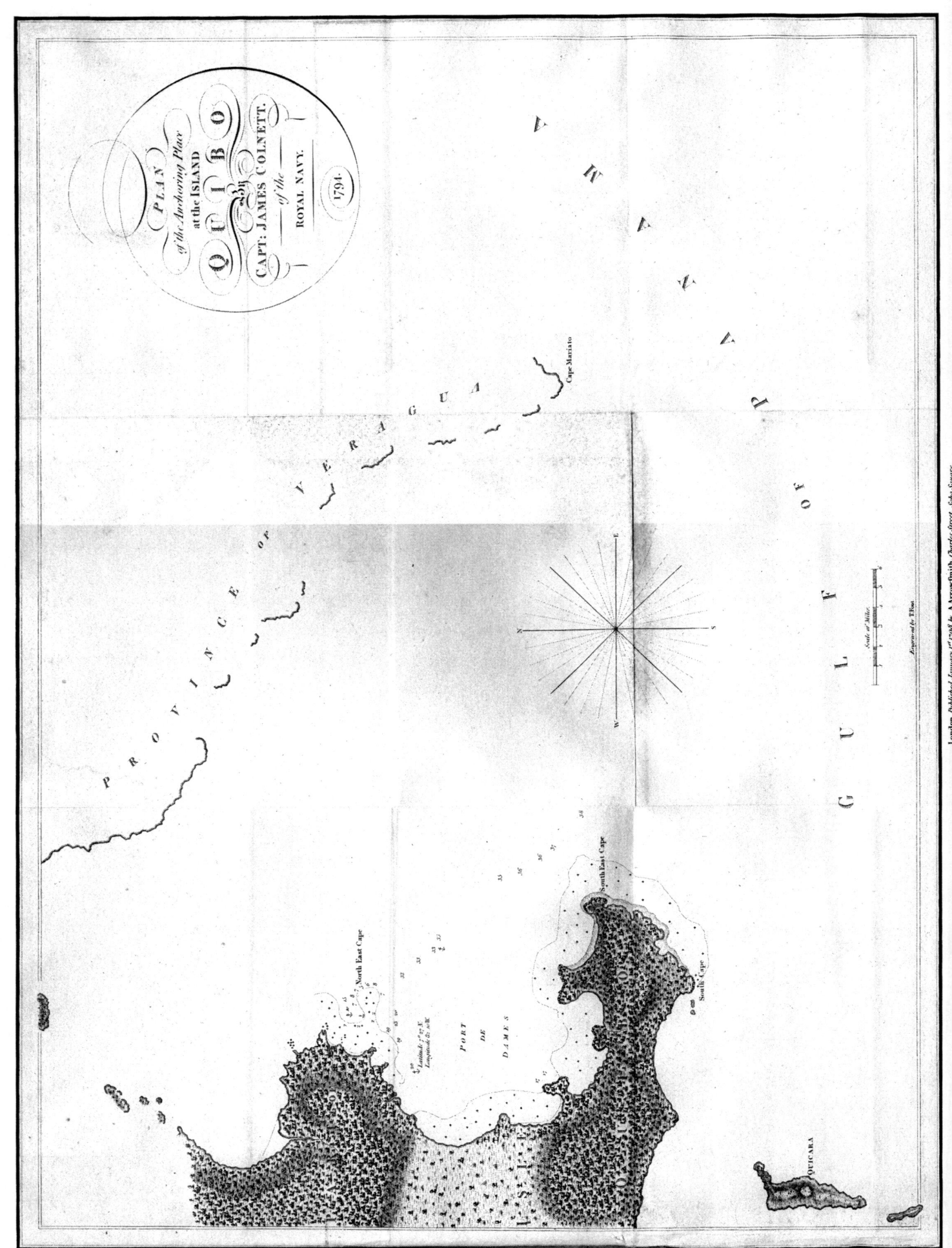

The material originally positioned here is too large for reproduction in this reissue. A PDF can be downloaded from the web address given on page iv of this book, by clicking on 'Resources Available'.

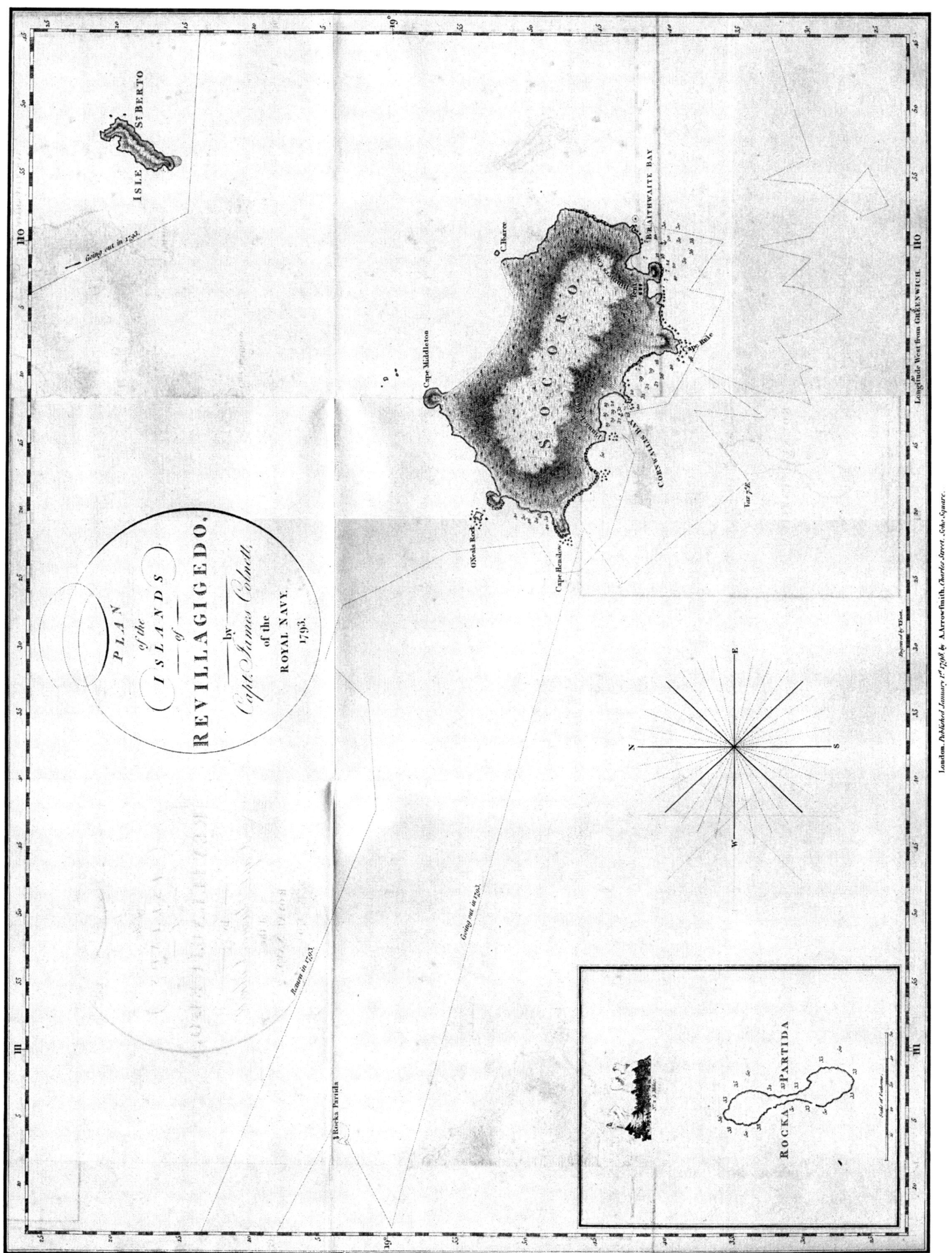

CHATHAM BAY in the Isle of COCAS, dist 3 or 4 miles

A. The Bay we left our Hogs & Goats in and Sow'd every kind of Garden seeds
B. Waters Bay just opening

S.E. Part of the Isle of SOCORA, taken at ½ dist from shore 1¾ miles

A. The E. Point the One our Tent was fixd in
B. The S Point the Isle

DIEGO RAMEREZ, 5 or 6 Leagues

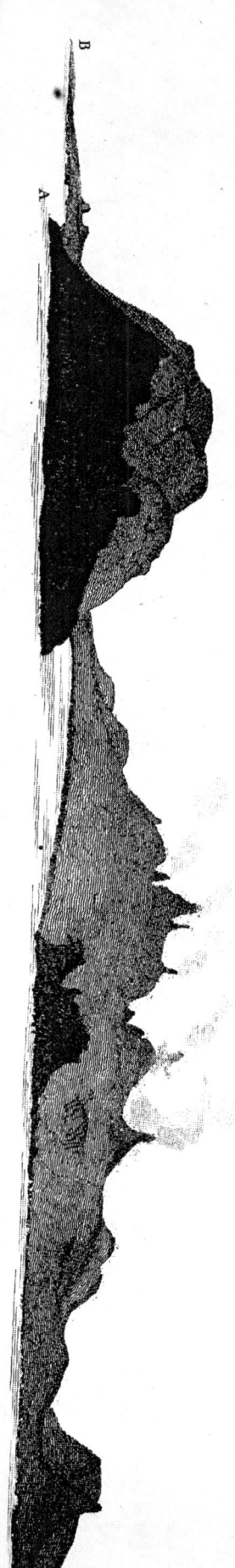

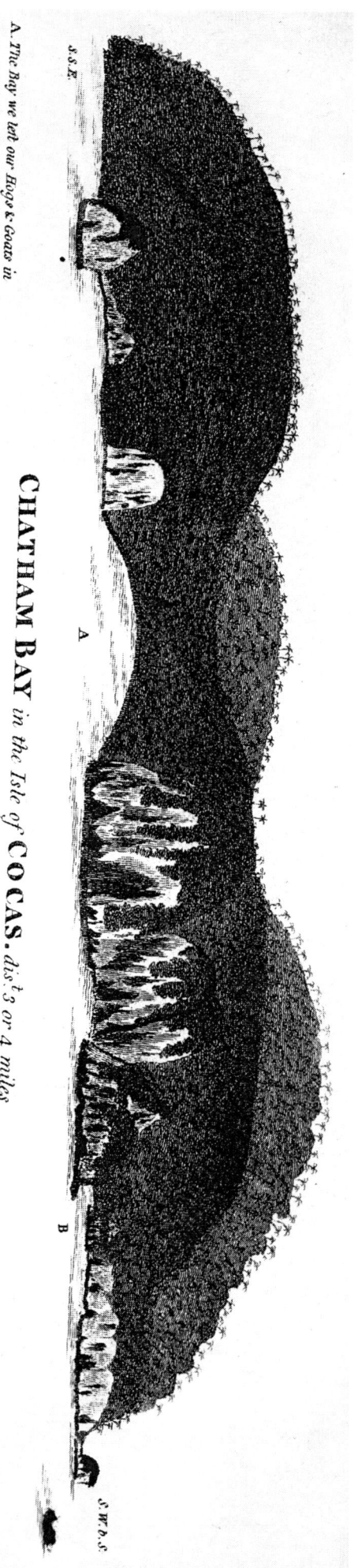

VIEW of the N.W end of JAMES'S ISLAND, one of the Galapagos taken at ‡.

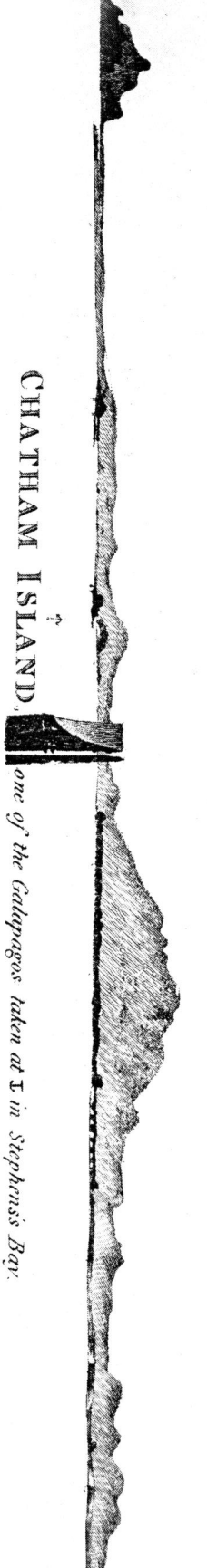

N.7.W.   A   B   C   ‡   D E.17.S.   E   S.24. F   G

A   Albanii Isle.
B   The N.W. point round which is a small Bay which I take for the one the Buccaneers call'd fresh water Bay in which were many of their Trees such as old Jars &c. also ground cleared away either as a Platform for Guns or to land stores &c. but the water since then has taken a different course & falls down between two hills at C & runs over a little dirt or Rocks into the sea.
D   Where we wooded & hauled the Sein & caught plenty of fish.
E   A small beach where the Buccaneers had Indian made their general landing Place & have raised Benches to sit on from here you may walk 10 miles in a Delightfull grove.
F   Cowley's enchanted Isle.
G   Part of Albemarle Isle.

CHATHAM ISLAND, one of the Galapagos taken at I in Stephens's Bay.

CHATHAM ISLE, one of the Galapagos taken at ‡ 1½ miles dist.

F.34.N.   S.45.W.   W.

PHYSETER, OR SPERMACETI WHALE.

Drawn by Scale, from one killed on the Coast of Mexico,

August 1793. and hoisted in on Deck.

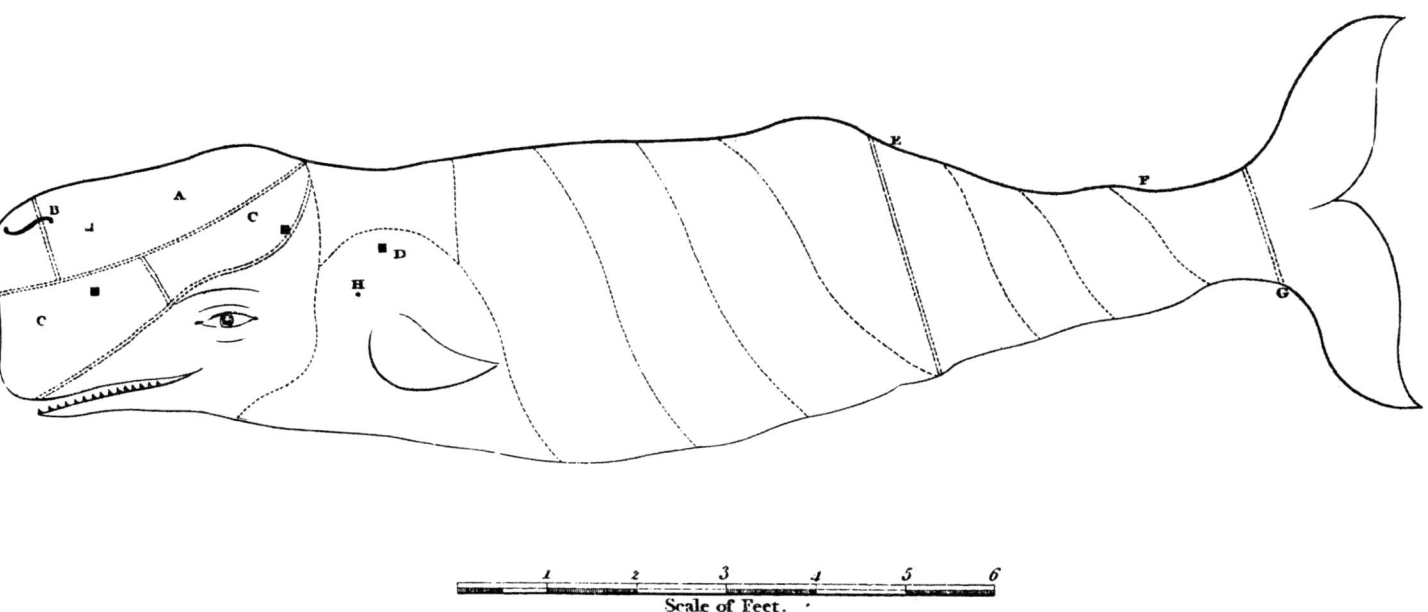

Scale of Feet.

A. Part of the Head containing liquid Oil, which is covered with a black membrane. B. The Spout-hole which runs horizontally along the left side, and is also seperated by the same kind of membrane. The part between the two double lines, is cover'd with Fat of considerable thickness, like that of a hog, these parts make one third of the quantity of Oil the Fish produces, of which the liquid is about one third. A.B. Part of the Head which of large Whales being too bulky and ponderous to be hoisted on board, is suspended in tackles and the front part cut off as described thus, and the Oil bailed out with buckets; but in small Whales, the head is divided at the double line below C.C. and hoisted upon deck. ▪▪ Where the tackles are toggled or hook'd. D Where the tackles are first hooked, which is called raising a peice, being thus steadied in the tackles the head is divided at the lowest double line and wore a stern till the fish is flinched, which is done by seperating the Fat from the Body with long-handled Iron Spades, as the Whale is hove round by the tackles the Fat peels off, and if any Sea is on the rising of the Ship considerably expedites the business. E. A large lump of Fat. F. A smaller. when the Fish is flinched, or peeled to E. it will no longer cant in the tackles, is therefore cut through at the first double line and also at G. the Tail being of no value. H. The Ear, which is remarkably small in proportion to the body, as is also the Eye from which a hollow or concave line runs to the forepart of the head the Eyes being prominent enables them to pursue their Prey in a direct line, and by inclining the head a little either to the right or left to see their enemy a stern, they have only one row of Teeth, which are in the lower Jaw with sockets in the upper one to receive them, the number depends on the age of the Fish, the lower Jaw is a solid Bone that narrows nearly to a point and closes under the upper, when they spout, they throw the water forwards and not upwards like other Whales except when they are enraged, they also spout more regular and stay longer under water the larger the Fish the more frequently they spout and continue longer under water. The Tail is horizontal with which he does much mischief in defending himself. Their Food, from all the observations I have had an oppertunity of making, has been the Sepia or middle Cuttle Fish. This species of the Whale, is remarkable for its attachment and for afsisting each other when struck with a harpoon: and more mischief is done by the loose Fish, than those the boats are fast to, and they frequently bite the lines in two which the struck Fish is fast with. The Ambergrease is generally discover'd by probing the intestines with a long Pole, when the Fish is cut in two at E.

Lightning Source UK Ltd.
Milton Keynes UK
UKOW020031150512

192562UK00001B/2/P